Sven Krueger · Wolfgang Gessner (Eds.)//
Advanced Microsystems for Automotive Applications
Yearbook 2002

Springer
Berlin
Heidelberg
New York
Barcelona
Hong Kong
London
Milan
Paris
Tokyo

Sven Krueger · Wolfgang Gessner (Eds.)

Advanced Microsystems for Automotive Applications
Yearbook 2002

With 193 Figures

 Springer

SVEN KRUEGER
VDI/VDE-Technologiezentrum Informationstechnik GmbH
Rheinstr. 10B
D-14513 Teltow
e-mail: krueger@vdivde-it.de

WOLFGANG GESSNER
VDI/VDE-Technologiezentrum Informationstechnik GmbH
Rheinstr. 10B
D-14513 Teltow
e-mail: gessner@vdivde-it.de

ISBN 3-540-43232-9 Springer-Verlag Berlin Heidelberg NewYork

Cip data applied for

This work is subject to copyright. All rights are reserved, whether the whole or part of the material is concerned, specifically the rights of translation, reprinting, reuse of illustrations, recitation, broadcasting, reproduction on microfilm or in other ways, and storage in data banks. Duplication of this publication or parts thereof is permitted only under the provisions of the German Copyright Law of September 9, 1965, in its current version, and permission for use must always be obtained from Springer-Verlag. Violations are liable for prosecution act under German Copyright Law.

Springer-Verlag is a company in the BertelsmannSpringer publishing group
http://www.springer.de
© Springer-Verlag Berlin Heidelberg New York 2002
Printed in Germany

The use of general descriptive names, registered names, trademarks, etc. in this publication does not imply, even in the absence of a specific statement, that such names are exempt from the relevant protective laws and regulations and therefore free for general use.

Typesetting: Camera ready by authors
Cover-Design: de´blik, Berlin
Printed on acid free paper   SPIN: 10865703   68/3020/kk - 5 4 3 2 1 0

Preface

In recent years microsystems applications (MST) in automobiles have become commonplace: they enabled the introduction of a series of new functions and at the same time the replacement of existing technologies with MST-based devices which offered improved performance and better value for money. In spite of the enormous progress to date, the results achieved reveal only the beginning of a revolution in the vehicle sector, which implies a complete transition from the mechanically-driven automobile system to a mechanically-based but ICT-driven system as part of a likewise complex environment.

EUCAR, the European Council for Automotive R&D, has defined three major strategic R&D areas:
- environment, energy, resources: clean, silent and lean transportation,
- road safety with the ultimate goal of a zero accident rate, and
- a sustainable and efficient traffic and transport system providing mobility and goods delivery services to the citizen.

Microsystems are indispensable for fulfilling these ambitions; they are often the pacemakers for such developments. At present highest importance is attributed to microsystems applications in safety, followed by engine and power train applications and on-board vehicle diagnostics.

The Yearbook published on occasion of the *2002 Conference on "Advanced Microsystems for Automotive Applications"* provides information on the state-of-the-art of R&D as well as existing technologies and potential developments in the sector. It is based on the conference contributions but goes further.

It gives both an overview for systems designers and, more generally, raises awareness of the strengths and opportunities offered by microsystems technology solutions. Wherever possible, other strategic, structural and market-related issues are dealt with, in addition to the technological aspects.

Since it will never be possible to be complete, the Yearbook aims to provide an annual forum through which to express the priorities of the automotive industry itself. The 2002 Yearbook "Advanced Microsystems for Automotive Applications" gives an overview on the state-of-the-art of microsystems technology for automotives including technology trends and market perspectives. Particular attention is focussed on engine and power train applications as well as obstacle detection which were also selected as priority themes for the 6th Conference on Advanced Microsystems for Automotive Applications.

I would like to express my sincere thanks to the authors for their valuable contributions to this publication and to the members of the Honorary and Steering Committee for their commitment and support. My explicit thanks go to Thomas Görnig, Conti Temic, Roger Grace, Roger Grace Associates, Manfred Klein, DaimlerChrysler and Detlef Ricken, Delphi Delco for the inspiring conversations we had. Special thanks are addressed to Florian Solzbacher, First Sensor Technology, to whom I am particularly indebted for his engagement in realising this publication.

Last but not least, I would like to express my thanks to the Innovation Relay Centre team at VDI/VDE-IT, Andrea Thiel for preparing the book for publication, and not forgetting Sven Krüger, project manager and responsible for this initiative.

Teltow/Berlin, March 21, 2002

Wolfgang Gessner

Public Financiers

Berlin Senate for Economics and Technology
European Commission
Ministry for Economics Brandenburg

Supporting Organisations

A.D.C.
Conti Temic microelectronic GmbH
DaimlerChrysler AG
Deutsche Bank AG
First Sensor Technology GmbH
Innovation Relay Centre Automotive Group
Investitionsbank Berlin (IBB)
mst*news*
Raytheon Commercial Infrared

AMAA Honorary Committee

Domenico Bordone	President and CEO
	Magneti Marelli S.p.A., I
Guenter Hertel	Vice President Research and Technology
	DaimlerChrysler AG, D
Rémi Kaiser	Director Technology and Quality
	Delphi Automotive Systems Europe, F
Hajime Kawasaki	Senior Vice President
	Nissan Motor Co. Ltd., J
Gian C. Michellone	President and CEO
	Centro Ricerche FIAT, I
Karl-Thomas Neumann	Head of Electronic Strategy
	VW AG, D

AMAA Steering Committee

Giancarlo Alessandretti	Centro Ricerche FIAT, I
Wilhelm Bois	Audi AG, D
Serge Boverie	Siemens VDO Automotive, F
Albert Engelhardt	Conti Temic microelectronic GmbH, D
Roger Grace	Roger Grace Associates, USA
Henrik Jakobsen	SensoNor A.S., N
Sven Krueger	VDI/VDE-IT, D
Peter Lidén	AB Volvo, S
Ulf Meriheinä	VTI Hamlin, FIN
Roland Mueller-Fiedler	Robert Bosch GmbH, D
Paul Mulvanny	QinetiQ, UK
Andy Noble	Ricardo Consulting Engineers Ltd., UK
Gerhard Pellischek	CLEPA, B
David B. Rich	Delphi Delco Electronics Systems, USA
Detlef E. Ricken	Delphi Delco Electronics Europe GmbH, D
Jean-Paul Rouet	SAGEM SA, F
Christian Rousseau	Renault S.A., F
Ernst Schmidt	BMW AG, D
John P. Schuster	Motorola Inc., USA
Bob Sulouff	Analog Devices Inc., USA
Egon Vetter	EWV Management Consultancy Pty Ltd., AUS
Ralf Voss	DaimlerChrysler Corp., USA
Matthias Werner	Deutsche Bank AG, D

Chairman:

Wolfgang Gessner	VDI/VDE-IT, D

Table of Contents

1	**Microsystems for Automotive Applications: Markets and Perspectives**	1
1.1	Markets and Opportunities for MEMS/MST in Automotive Applications R. Grace	3
1.2	The Automotive Microsystems Market: Insights from the NEXUS Market Study 20012 H. Wicht, M. Illing, R. Wechsung	13
1.3	Automotive Micro and Nano- Technology in the European Sixth Framework Programme T. Reibe	21
1.4	Regulations for Automotive Applications T. Goernig, S. Krueger	26
2	**Microsystems Technologies and Materials**	29
2.1	Microsystems Technology – a New Era Leading to Unequalled Potentials for Automotive Applications F. Solzbacher	31
2.2	Fabrication Technologies for 3D-microsensor Structures F. Solzbacher	35
2.3	Signal Processing for Automotive Applications T. Riepl, S. Bolz, G. Lugert	40
2.4	Materials for Electronic Systems in Automotive Applications - Silicon On Insulator (SOI) P. Seegebrecht	54
2.5	Microsystems Packaging for Automotive Applications E. Jung, M. Wiemer, V. Grosser, R. Aschenbrenner, H. Reichl	66
2.6	Micro-Mechatronics in Automotive Applications F. Ansorge, J. Wolter, H. Hanisch, V. Grosser, B. Michel, H. Reichl	76

3 Functions and Applications — 87

3.1 Towards Ambient Intelligence - Obstacle Detection — 89

3.1.1 Overview: Obstacle Detection — 89
A. S. Teuner

3.1.2 Environment Sensing for Advanced Driver Assistance – CARSENSE — 96
J. Langheim

3.1.3 True 360° Sensing Using Multiple Systems — 106
T. Goernig

3.1.4 Electronic Scanning Antenna for Autonomous Cruise Control Applications — 116
B. Kumar, H.-O. Ruoß

3.1.5 Advanced Uncooled Infrared System Electronics — 121
H. W. Neal, C. Buettner

3.1.6 Laserscanner for Obstacle Detection — 136
U. Lages

3.1.7 Infrared Camera for cAR – ICAR a EURIMUS Project for Driver Vision Enhancement — 141
J.P. Chatard, X. Zhang, E. Mottin, J.L. Tissot

3.1.8 Micro-Opto-Electro-Mechanical-Ssystems (MOEMS) in Automotive Applications — 146
F. Ansorge, J. Wolter, H. Harnisch, H. Reichl

3.2 Powertrain Applications — 157

3.2.1 Overview:
Microsystems as a Key Eelement for Advanced Powertrain Systems — 157
A. Noble, M. Voigt, S. Krueger

3.2.2 A Piezoelectric Driven Microinjector for DI-Gasoline Applications — 164
C. Anzinger, U. Schmid, G. Kroetz, M. Klein

3.2.3 A Robust Hot Film Anemometer for Injection Quantity Measurements — 174
U. Schmid, G. Kroetz

3.2.4	**Non-Contact Magnetostrictive Torque Sensor – Opportunities and Realisation** C. Wallin, L. Gustavsson	184
3.2.5	**Benefits of a Cylinder Pressure Based Engine Management System on a Vehicle** P. Moulin, R. Mueller, J. von Berg et al.	196
3.3	**Measurement - Angle and Position Sensors for Vehicles**	206
3.3.1	**TPO (True Power On) Active Camshaft Sensor for Low Emission Regulation** G. Carta	206
3.3.2	**Contactless Position Sensors for Safety-critical Applications** Y. Dordet, B. Legrand, J.-Y. Voyant, J.-P. Yonnet	212
3.3.3	**High Resolution Absolute Angular Position Detection with Single Chip Capability** H. Grueger, H. Lakner	222
3.3.4	**Programmable Linear Magnetic Hall-Effect Sensor with Excellent Accuracy** S. Reischl, U. Ausserlechner	227
3.4	**Measurement - Pressure Sensing in Automotive Applications**	232
3.4.1	**Overview: Principles and Technologies for Pressure Sensors for Automotive Applications** C. Cavalloni, J. von Berg	232
3.4.2	**Tire Pressure Monitoring Systems – the New MEMS Based Safety Issue** J. Becker	243
3.4.3	**Measurement of the Actuation Force of the Brake Pedal in Brake-by-wire Systems Via the Use of Micromechanical Sensors** E. Weiss, G. Braun, E. Schmidt, M. Dullinger, U. Witzel	251
3.4.4	**Design of a New Concept Pressure Sensor for X-by-wire Automotive Applications** L. Tomasi et al.	261

3.4.5	**Pressure Sensors in the Pressure Range 0 — 300 bar for Oil Pressure Applications - Directly in the Oil Media** R. Schmidt	271
3.5	**Further Functions and Applications**	276
3.5.1	**Sensors in the Next Generation Automotive Networks** G.H. Teepe	276
3.5.2	**Plastic Packaging for Various Sensor Applications in the Automotive Industry** I. van Dommelen	289
3.5.3	**Harsh Fluid Resistant Silicone Encapsulants for the Automotive Industry** K.L. Pearce, E.M. Walker, J. Luo, R.A. Schultz	297
3.5.4	**Silicon-On-Insulator (SOI) Solutions for Automotive Applications** L. Demeûs, P. Delatte, G. Picun, V. Dessard, Pr.D. Flandre	306
3.5.5	**Micromachined Gyroscope in Silicon-On-Insulator (SOI) Technology** A. Gaisser et al.	311
3.5.6	**Thermal Microsensor for the Detection of Side-Impacts in Vehicles** M. Arndt, R. Aidam	316
3.5.7	**Low-cost Filling-level Sensor** H. Ploechinger	321
3.5.8	**IBA - Integrated Bus Compatible Initiator** T. Goernig	325
Contributors		331
Keywords		335

1 Microsystems for Automotive Applications: Markets and Perspectives

1.1 Markets and Opportunities for MEMS/MST in Automotive Applications

R. H. Grace

Roger Grace Associates
83 Hill Street, San Francisco, CA 94110, USA
Phone: +1/415/436-9101, Fax +1/415/436-9810
E-mail: rgrace@rgrace.com

Keywords: challenges, application status, market analysis, products, trends

Abstract

This paper will focus on the evolution of and on the current and future automotive applications of MEMS (Microelectro-mechanical Systems), and MST (Microsystem Technologies). Current product developments in pressure sensors, accelerometers, angular rate sensors, and other MEMS/MST devices are presented. Market figures for automotive electronics and automotive sensors from 2001-2008 are given.

Introduction

Microelectromechanical Systems (MEMS) and Microsystem Technologies (MST) have played a role in the automotive industry for over 20 years. The first MEMS/MST devices were manifested as manifold absolute pressure (MAP) sensors used in the electronic engine control system (EECS). Here they measure the manifold air pressure which was used as an inferential means of determining the air/fuel mixture entering the combustion cylinder. Judicious control of this function maximised fuel economy and minimised uncombusted hydrocarbons. This solution was basically mandated by Federal government requirements. MAP, and their barometric pressure cousins (BAP) are still used in large volumes in today's vehicles. Over 50 million of these devices were sold in 2001. The motivation to use sensors, especially MEMS/MST-based ones are being driven by the following factors:

- Longer warranty periods (up to 10 years/150,000 miles) require increased reliability components
- Continuously developing federal and regional mandated fuel efficiency, emission and safety standards.

- Higher vehicle performance and comfort
- Availability of low-cost microcontrollers, memory and displays
- Enhanced vehicle diagnostics

Market Figures

World-wide vehicle production has grown from 51.8 million passenger vehicles, SUVs and light trucks in 2001 to a projected 58.5 million vehicles by the year 2008. The 2001 market for automotive electronics was $25.7 Billion (US) growing to an expected $42.4 Billion (US) in 2008, representing a compounded annual growth rate of 6.6%. This equates to an average electronic content per vehicle of $497 (US) in 2001 to $725 (US) in 2008.

It has been reported that the automotive sensor market to be worth $6.4 Billion (US) in 2001 and will grow at a compounded annual growth rate (CAGR) of 5.7% in dollars and 8.2% in units to $9.4 Billion (US) (1,750 billion units) by 2008. North America accounts for 44% of the year 2001 market for automotive sensors, followed by Europe (28%), Japan, (23%), and then South Korea (5%) [1].

Some of the more high compounded annual growth rate (CAGR) sensors in the 1999-2008 period include: adaptive cruise control distance sensors (65.4%), air quality sensors (46.2%), occupancy detection (39.5%), torque sensing (27.0%), parking aid (27.1%), load sensing (20.9%), and pressure sensing (14.0%). Market figures for 2008 in $US show the largest dollar volume opportunities to be: position sensing ($19.6B), pressure sensing ($16.4B), speed sensing ($19.6B), pressure sensing $16.4B), speed sensing ($15.0B), and oxygen sensing ($13.1B) [1].

MEMS/MST technologies are currently satisfying or have a great opportunity to fill many of these applications and are expected to constitute a greater share of the automotive sensor market in 2008 versus its current value. A recent Roger Grace Associates/Nexus study has estimated that the 2000 sales of Automotive MEMS/MST to grow from $1.75 Billion (US) to $2.27 Billion (US) by 2005, which constitutes a 16.9% compound annual growth rate. The total MEMS/MST market is estimated to grow from $14.11 Billion (US) in 2000 to $36.22 Billion (US) by 2005, constituting a CAGR of 20.1% [3].

System Applications

We have reviewed the use of MEMS/MST in the following automotive systems:

- Safety
- Engine/Drive Train
- Comfort, Convenience, Security, Engine/ Drive Train
- Vehicle Diagnostics/Monitoring

For each system, specific applications are noted (e.g., Digital Engine Control Fuel Level) and the status of the specific application is given (i.e., future, limited production, major production). Also noted is the opportunity afforded to a MEMS/MST solution versus that using another technology (e.g., piezoelectric, hall effect). The appropriateness criteria for MEMS/MST selection was based on detailed market research initially conducted by the author and first reported in 1991 and updated periodically and as recently as February, 2001 during the annual SAE Conference in Detroit, Michigan. The most significant application opportunities will be addressed.

Safety

A summary of MEMS/MST safety application opportunities in automotive safety systems is given in Table 1.

Airbag actuation is currently and will continue to be a major application of MEMS. Silicon accelerometers in the 50g range are currently being supplied to U.S. vehicles, primarily by Analog Devices, Motorola, Bosch, Sensonor (Norway), and Nippon Denso (Japan).

Three new applications in the safety sector warrant some comments. Yaw Rate sensors (angular rate sensors) have been developed by a number of manufacturers including Systron Donner, Bosch, Nippon Denso, and Analog Devices. This is a relatively new application with Systron Donner being the only high-volume supplier to date. Here, the angular rate sensor is used in a vehicle dynamic control system (VDC) along with a suite of other sensors to assist in the control vehicle stability in high-speed turning/cornering scenarios. In addition, yaw rate sensors are being used in vehicle roll-ver applications (e.g. Bosch).

Application	Sensor/Structure	Status	MEMS/MST Opportunity
Antilock Braking/Vehicle Dynamics/Suspension			
	Steering Position	Production	Medium
	Wheel Rotation	Production	Medium
	Pressure	Limited Prod.	Medium
	Acceleration	Limited Prod.	High
	Valve	Future	Low
	Rate	Limited Prod.	High
	Displacement	Limited Prod.	Low
	Rollover	Limited Prod.	High
Airbag Actuation			
	Acceleration	Production	High
	Pressure (Canister)	Future	Medium
	Pressure (Side Impct)	Limited Prod.	High
Seat Occupancy/Passenger Position			
	Presence/Force	Limited Prod.	High
	Displacement	Limited Prod.	High
Seatbelt Tensioner			
	Acceleration	Limited Prod.	High
Object Avoidance			
	Presence/Displcmnt.	Limited Prod.	High
Navigation			
	Yaw Rate/Gyro	Limited Prod.	High
	Wheel Rotation	Limited Prod.	High
Road Condition			
	Optical	Future	High
Headlight Leveling			
	Tilt	Limited Prod.	High
Adaptive Cruise Control			
	RF MEMS	Future	Medium
	Displacement	Limited Prod.	Medium

Table 1. Applications of MEMS/MST: Safety Group

MEMS/MST-based solutions are the enabling technology for passenger safety applications. To date, the automotive manufacturers or their first tier suppliers have not agreed upon a standard solution approach; and most

1.1 Markets and Opportunities for MEMS/MST in Automotive Applications

experts feel that they never will. As a result, various approaches using totally different MEMS/MST (as well as non-MEMS/MST) solutions are under development. Some of the more interesting solutions use a myriad of sensors to optimise the efficacy of the system. Among these are infrared proximity sensor arrays to determine the location of the passenger in 3-dimensional space; force sensors in seats to weigh the passenger, and/or to determine if a passenger is occupying the seat; pressure sensors in bladders embedded in the car seat to measure passenger presence/weight; force sensing resistor arrays to determine passenger weight and location on the seat. It should be quite obvious that this is a complex problem that will take a sophisticated system of hardware and software algorithms to optimally solve the problem.

Comfort, Convenience and Security

A summary of MEMS/MST application opportunities in automotive comfort, convenience and security systems is given in Table 2.

Application	Sensor/Structure	Status	MEMS/MST Opportunity
Seat Control			
	Presence	Limited Prod.	Low
	Valve	Future	Low
	Displacement	Future	Low
Climate			
	Mass Air Flow	Future	Medium
	Temperature	Production	Medium
	Humidity	Limited Prod.	High
	Air Quality	Limited Prod.	High
Compressor Control			
	Pressure	Production	High
	Temperature	Production	Low
Security			
	Proximity	Limited Prod.	Low
	Motion	Limited Prod.	Medium
	Vibration	Limited Prod.	Medium
	Displacement	Limited Prod.	Low
	Fingerprint	Limited Prod.	Medium
Windshield Wipers			
	Optical	Limited Prod.	Medium
	Optical	Future	Medium

Table 2. Applications of MEMS/MST:Comfort, Convenience, and Security Group

The measurement of compressor pressure in the vehicle air conditioning system offers a major opportunity for MEMS/MST. Currently, other technologies (e.g., Texas Instrument ceramic capacitive pressure sensor) are being used. Major developments by a number of MEMS/MST companies are actively pursuing this very large opportunity, including Measurement Specialities which has recently formed a strategic alliance with Texas Instruments. Keller, Fasco, and ISS/Texas Instruments are also pursuing this opportunity.

Engine/Drive Train

A summary of MEMS/MST application opportunities in automotive engine/drive train systems is given in Table 3.

Application	Sensor/Structure	Status	MEMS/MST Opportunity
Digital Engine Control			
Fuel	Level	Production	Low
Cylinder	Pressure	Future	Medium
Manifold (MAP)	Pressure	Production	High
Barometric	Pressure	Production	High
Eng Knock	Vibration	Production	Medium
Mass Airflow	Flow	Limited Prod.	Medium
Exhaust	Gas Analysis	Production	Low
Crankshaft	Position	Major Prod.	Medium
Camshaft	Position	Limited Prod.	Medium
Throttle	Position	Limited Prod.	Medium
EGR	Pressure	Production	High
Fuel Pump	Pressure	Future	High
Torque	Torque	Limited Prod.	Medium
Continuously Variable Transmission			
	Temperature	Future	Low
	Pressure	Limited Prod.	High
	Microvalve	Future	Low
Fuel Injection			
	Pressure	Limited Prod.	High
	Nozzle	Limited Prod.	High
Diesel TurboBoost			
	Pressure	Limited Prod.	High

Table 3. Applications of MEMS/MST: Engine/Drive Train

Electronic engine control has historically been and is expected to be a major application area of MEMS/MST in automotive applications. Silicon manifold absolute pressure (MAP) sensors are produced by the millions by Delco, Motorola, and Bosch. These devices provide an inferred value of air-to-fuel ratio by measuring intake manifold pressure. A great deal of

1.1 Markets and Opportunities for MEMS/MST in Automotive Applications

effort has been undertaken to replace these devices with mass airflow (MAF) devices. Currently available on the market are discrete hot wire anemometer devices (e.g., Hitachi). Because of their construction, they tend to be large and expensive. A thin film equivalent of this device was introduced by Bosch in 1995. A MEMS/MST version of this device is currently under evaluation by a number of organisations. In addition to the MAP/MAF devices, barometric pressure values are needed to provide the engine controller with altitude information to compensate a rich/lean fuel-to-air mixture. MEMS/MST devices are well suited for this application.

Cylinder pressure values are of great importance to optimise engine performance; however, due to the extreme high temperature levels, piezoelectric and fiber-optic techniques such as that developed by Optran provide a much more pragmatic solution to this application; however, at this time cost issues preempt their introduction. Exhaust gas recirculation (EGR) applications exist in Ford and Chrysler systems. Ceramic capacative pressure sensors are being replaced by silicon piezoresistive solutions.

Continuously variable transmission (CVT) applications require pressure measurements in hydraulic fluids. MEMS/MST devices which are isolated from the media using various techniques (e.g., isolated diaphragms) could find widespread application. Fasco, Measurement Specialties, Integrated Sensor Solutions (recently acquired by Texas Instruments), and SSI Technologies are developing possible solutions to this application. All of these approaches use a sensor plus silicon CMOS ASIC hybrid, typically electronically programmed using EEPROMS.

The only known application of a MEMS device in a mechanical structure is in fuel injector nozzles. Here, Ford has used micromachined silicon to create highly uniform and rectangular orifices for fuel injection systems. Over 3 million of these devices were manufactured; however, they are not currently in production.

Vehicle Diagnostics/Monitoring

A summary of MEMS application opportunities in automotive vehicle diagnostic/ monitoring is given in Table 4.

One of the more interesting applications for MEMS/MST is in tire pressure monitoring. For both safety and optimised fuel performance, proper tire inflation is necessary. A number of systems are currently being offered that provide real time measurement of tire pressure. MEMS devices are ideally suited and are being considered by a number of their manufacturers for this purpose. With the favourable acceptance of run-flat tires, these systems became very popular by the model year 2000 vehicles. Run-flat tires (e.g., Michelin 60-series) eliminate the cost and weight of a spare and jack. TRW, NovaSensor, Motorola, and Sensonor are currently pursuing this application, which has become a major opportunity as a result of the recent Clinton Administration's edict that all passenger vehicles operating in the US be equipped with these devices by 2004.

Application	Sensor/Structure	Status	MEMS/MST Opportunity
Coolant System			
	Temperature	Production	Low
	Level	Limited Prod.	Low
Tire			
	Pressure	Limited Prod.	High
Engine Oil			
	Pressure	Production	High
	Level	Production	Low
	Contamination	Limited Prod.	Medium
Brake System			
	Pressure	Limited Prod.	High
	Level	Future	Low
Transmission Fluid			
	Pressure	Limited Prod.	High
	Level	Future	Low
Fuel System			
	Pressure	Limited Prod.	High
	Level	Future	Low
	Pressure (Vapor)	Limited Prod.	High
Vehicle Speed			
	Velocity	Production	Medium

Table 4. Applications of MEMS/MST: Diagnostics/Monitoring

Engine oil monitoring is a huge opportunity for MEMS/MST. The greatest barrier to the adoption of these systems is price. These pressure sensors must be able to survive the elevated temperature requirements of engine oil and isolate the silicon chip from the media. Price target for this application is in the $5–$7 range for a fully signal conditioned, packaged device. Numerous sensor manufacturers are aggressively pursuing this significant application opportunity.

Recent legislation has created a major opportunity for pressure sensors in the evaporative fuel system. In this application, a pressure sensor is used to monitor the pressure level in the fuel tank and insure that no fuel vapor escapes.

Conclusions

The automotive industry was one of the earlier adopters of MEMS/MST behind the military and industrial control sectors. The market holds great promise for MEMS/MST companies wishing to participate because of the large number of current and even larger number of ever increasing new applications. Many older technologies that have secured a position into an automotive system are being replaced at a rapid rate by MEMS/MST solutions. The automotive sensor market is expected to grow at 5.2% CAGR from $6.103 billion (US) in 1999 to $9.361 billion (US) in 2008. This growth rate is significantly less than the overall MEMS growth rate of 21%.

Major opportunities for high growth opportunities exist in adaptive cruise control distance sensors, air quality sensors, torque sensors, parking aid sensors, load sensors and pressure sensors.

Companies wishing to be successful with automotive manufacturers and tier one suppliers must provide the value with their product offerings. These customers are buyers of solutions and not technology.

The major challenges for MEMS/MST suppliers will continue to be low cost packaging and testing of sensor elements rather than the development of new sensor technology. There are currently MEMS/MST-based solutions for many applications. However, the long design-in-cycle and extended period to achieve full scale production status is the major reason for the low 5.2% CAGR. Another important issue is that suppliers are under contract pressure to reduce their prices. This is evident by the fact that sensor unit volume is expected to increase with a CAGR of 8.2% versus 5.2% in dollars volume over the 1999-2008 time future.

Acknowledgement

The author would like to thank Mr. Joe Giachino at the University of Michigan, Mr. Alby Berman of Tekata, Mr. Bob Sulouff for their helpful comments, and Mr. Chris Webber of Strategy Analytics for his valuable contribution of automotive sensor market data.

References

[1] C. Webber, "Automotive Sensor Market Trends." Presented at 1999 Sensors Marketing Conference, Strategy Analytics; San Diego, CA; Feb. 1999.

[2] R.H. Grace, P. Salomon, "Microsystems, MEMS, Micromachines: On the Move from Technology to Business." MST News; May 2, 2001; pp. 4-7.

[3] S. Krueger, R.H. Grace, "New Challenges For Microsystems Technology In Automotive Applications." MST News; January, 2001; pp. 4-8.

1.2 The Automotive Microsystems Market: Insights from the NEXUS Market Study 2001

H. Wicht[1], M. Illing[2], R. Wechsung[3]

[1]WTC-Wicht Technologie Consulting München
 E-mail: henning.wicht@wtc-consult.de

[2]Robert Bosch GmbH Reutlingen

[3]Steag Microparts GmbH Dortmund

Keywords: automotive trends, market analysis, NEXUS, products

Abstract

This paper summarizes the findings of the NEXUS market study 2001 and focuses on products, applications and trends for MEMS in automotive. The NEXUS market report is based on a top down analysis of application fields and a bottom up analysis of products. Insights for automotive will start with the application fields followed by drivers and trends of innovation.

1 The MST market 2000 and 2005

The Nexus market analysis identified 17 MST and microstructured products. In 2000, the total turnover is estimated at 26 Bio US $ (Fig.1). It will increase to about 50 Bio US $ in 2005, which corresponds to 16% CAGR. The main products in 2000 are read-write heads for hard disk drives (HDD), inkjet printheads, Invitro Diagnostics and hearing aids, generating about 80% of the turnover. These products will continue to dominate the total market for MST in 2005, however the market share of HDD and Inkjet will decrease to some extent and In vitro Diagnostics will increase its share.

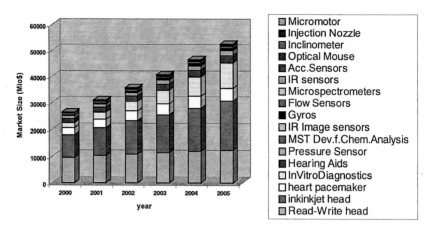

Fig. 1. MST Market Volume Existing Products (Source: Nexus 2002)

2 The Role of the Automotive Industry for MST

A large variety of the MST products have been applied primarily in the automobile. Pressure and acceleration sensors are well known examples and injection nozzles and inclinometers are especially developed for the automotive market. In addition Gyroscopes, IR and radar sensors will have a promising future in the car.

The automotive sector is looking for large volume series. That is the reason why MST offers Interesting solutions, even if products are under strong price competition.

While some MST markets - like - IT - are predominantly defined by the US and Asia, many of these MST innovations in the automotive field come from Europe, due to its strong user and supply base

3 The Automotive / Transportation Application Field

It is envisaged that future generations of cars will be even more fuel-efficient with higher safety and comfort standards. This will be enabled, primarily, through electronic systems. In 1999, the cost value of electronic components within a car was estimated to be in the order of 15% of the total cost. By the year 2003 this figure will, it is forecast, reach approximately 20% of the total cost. In this context, microsystems, and particularly micro-sensors, will play a major contributory role in the car of the future.

Automotive applications have been an early adopter and driver of microsystem technologies. This began in the 1980's with the development and use of micromachined silicon-based pressure sensors. Microsystems have, since then, been successfully utilised in many other automotive applications, primarily as sensors. Indeed, automotive sensors manufactured using microsystems technologies will, it is estimated, attract a 3.5 bn$ market by 2005 and constitute approximately 30% of the overall automotive sensor market. Growth rates for these microsystem applications are expected to be in excess of 20% per year.

The automotive applications demand high-quality products that must operate, flawlessly, under sometimes extreme conditions. Typical production volumes reach millions/year and the cost pressure is intense. These requirements are often met best by microsystem solutions, which explains their remarkable success in recent years for automotive applications.

3.1 Applications

The following Table 1 provides a list of the MST-based components and their respective applications within the automotive field:

Microsystem	Application	Technologies
Acceleration Sensor	Frontairbag, Sideairbag, Upfontsensors, Vehicle Dynamic Control, Active Suspension	Si Surface Micromachining, Si-Bulk Micromachining, ceramic piezoelectric
Angular Rate Sensors (Gyro, Yaw Rate)	Vehicle Dynamics Control, Rollover Sensing, GPS Navigation Systems	Si Surface Micromachining, Si-Bulk Micromachining, Quartz Micromachining
Pressure Sensors	Air intake, Atmospheric pressure, Fuel Vapour pressure, Turbocharger pressure, tire pressure, HVAC, Fuel injection	Si Bulk micromachining Si Surface micromachining
Flow sensors	Air intake	Si Bulk micromachining Hot wire
Chemical Sensors (electronic nose only)	HVAC	Si Bulk micromachining + Metaloxide-Thickfilm, Thinfilm
Nozzles	Fuel injection	Stamping, Laser, Micro-erosion in metal substrates
Fluid condition sensors*	Oil condition	Mostly conventional
Magnetic Sensors* for angle, position and speed.	Wheelspeed, Camshaft position and speed, pedal position, steering wheel angle	MR, Hall, inductive
Radar Sensors*	Parking aid, blind spot surveying, Cruise Control	Microwave sensor

* not considered as MST in this study

Table 1. Overview of Microsystems for Automotive Applications

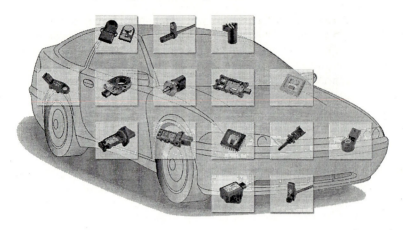

Fig. 2. Sensors are the main application field for microsystems in the car. A high-end car may contain between 50 and 100 sensors, many of which are MST-based. (Source: Bosch)

Components such as pressure and acceleration sensors have been used for many years for airbag safety systems and engine management. These applications continue to grow with new developments of accelerometers, yaw rate and roll-over sensors, child seat presence and occupant classification systems which will provide higher active and passive safety. In future, the airbag system will be able to sense the strength and direction of impacts more accurately and will trigger the enlargement of only half of the volume of the air-bag in the case of small accidents/impacts. Of late, microsystems have been introduced on the market for the continuous

Fig. 3. Silicon Micromachined Gyroscope for Navigation and Rollover-Protection (Source: Bosch)

remote measurement of tire temperature and pressure thus providing an additional safety feature. This market will be significantly strengthened by the year 2003 when the new US-legislation demanding tire pressure control in all cars is introduced.

In addition, new applications such as GPS (Global Positioning Systems) and HVAC (Heating, Ventilation Air Conditioning) for car safety and driver comfort systems are currently penetrating the market. These too make use of MST by integrating gyroscopes, pressure sensors and air quality sensors.

Engine optimisation, on the other hand, will be achieved through the use of pressure sensors, flow sensors, temperature sensors and rotary position sensors which will control the engine and provide data of the engine parameters to the driver.

3.2 Trends

The trends in the development of microsystems for automotive applications will address the continuing need for:

- increased safety
- increased comfort
- reduced power/energy consumption

The foreseen system trends are summarized in Fig. 4 below.

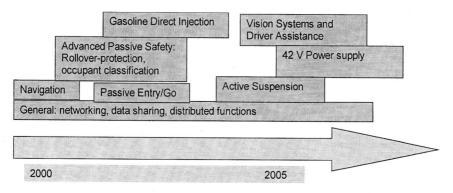

Fig. 4. Major System Trends Influencing Microsystems

At the microsystem component level the trends are shown in Fig. 5.

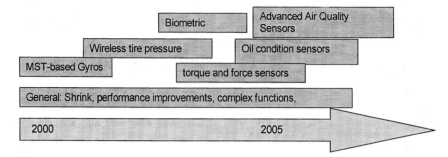

Fig. 5. Trends of Microsystems in Automotive

Well established microsystem-based devices for automotive applications will continue to gain market share replacing conventional technology. New systems will provide additional market opportunities: Examples include, increasing market penetration for the side-air-bag protection systems driving the accelerometer market and the requirement for (low-g) accelerometers will be stimulated by the need for improved active suspension systems and vehicle dynamics control systems. Pressure sensors are well established as manifold air pressure sensors, however, new growth market opportunities will arise from direct injection engines, HVAC and legislative requirements (e.g. fuel vapour pressure, tire pressure).

In addition, new microsystem-based sensors which will be commercialised for other applications may also find use within the automotive sector. Examples include the biometric sensor (used for fingerprint recognition) applied for "key-less-entry" of vehicles; Torque and force sensors, essential for the introduction of drive-by-wire systems; and advanced air quality sensors which will lead to improved HVAC systems as well as oil condition sensors allowing longer, oil change, intervals to be feasible.

Beyond 2005, radar and optical vision systems will, it is foreseen, form part of the next generation safety systems. It is not, as yet, clear whether the technology will be MST-based but aspects (subsystems) are likely to be as such (e.g. RF-MEMS, micro-optics).

3.3 Markets and Players

The drivers of innovation, namely safety, comfort, consumption and traffic management can be observed on a worldwide basis for the automotive business. The focus, however, is likely to be different between Europe, North America and the Far East.

In Europe, the MST gyroscope is forecast to become one of the most rapidly growing product for the future. This trend is driven by the need for higher stability control systems, improved restraint systems and novel navigation systems. Another European growth area relates to HVAC systems. The US and Asian markets are currently almost saturated, but the demand in Europe is still growing. The increase in the market size was estimated at 20 % in 1998. The demand for MST pressure sensors that regulate HVAC systems will, therefore, expand accordingly. Other trends include a continued, strong increase in the incorporation of side-airbag systems and direct injection diesel engines (fuel pressure sensors, atmospheric pressure).

The major MST-Manufacturers for automotive applications in Europe are Bosch, Siemens, Infineon, Conti-Temic, VTI-Hamlin, Philips, SensoNor and HL-Planartechnik

In the Far East and, particularly, Japan the major MST products for this market are acceleration sensors, gyroscopes as well as pressure and gas sensors. Navigation systems are the main driver for the use of gyroscopes where Japan takes a world-wide lead for such systems with equipment growth rates estimated to reach over 30% by 2003. Novel systems on the horizon include the Advanced Cruise-Assist Highway System (AHS), where MST components such as sensors, radar front-ends, CCD cameras and micro-lasers will be used for these systems.

The major MST-manufacturers for automotive applications, in Japan, are Denso, Matsushita, NGK, and Figaro.

Fig. 6. Tire Pressure Sensor (Source: Sensonor)

The North American market has been the main driver for improving passive safety and engine control systems for vehicles. Current major innovations, originating from the North American market, include passenger classification techniques and rollover-protection and tire pressure monitoring systems. In the future, pedestrian protection might form an integral part of such passive safety systems.

Further products, with strong growth potential on the US-market might be oil condition sensors, torque sensors and yaw rate sensors.

The major suppliers in the USA are Delphi, Motorola, Analog Devices, BEI Systems and Texas Instruments.

To conclude, the driving forces for MST in automotive are again safety and comfort. New mass products will enter the market e.g. gyroscopes and wireless tire pressure monitoring devices.

The automotive market is an established market, it means that the infrastructure to use MST products is established. Once the MST product has passed the qualification and is accepted, mass fabrication can be started. As in the past, automotive is a driver for innovations in MST. The NEXUS market study shows that this will continue in the next years.

Background Information

Nexus Market Analysis 2002

The Nexus Task Force Analysis is an indepeded group of experts evaluating market trends in MST/MEMS. Members are R. Wechsung, F. Goetz, H. Wicht, JC Eloy, A. El Fatatry, F. v.d. Weijer, G. Tschulena, M. Illing, R. Lawes, H. Zinner

The report "Market Analysis for Microsystems 2000 – 2005" can be ordered at www.wtc-consult.de.

1.3 Automotive Micro and Nano-Technology in the Sixth Framework Programme

T. Reibe, Scientific Officer European Commission

Information Society Directorate-General
Phone: +32.2.295.61.25
E-mail: thomas.reibe@cec.eu.int

1 Micro and Nano-Technology (MNT) in the Sixth Framework Programme (FP6)

1.1 Introduction

Micro and Nano-Technology is very wide ranging, covering a variety of disciplines, topics and activities at different stages of the development. 4 lines of activities (see Fig. 1) are of major concern in the next Framework (FP6) of European Research:

- micro- and nano-electronics oriented towards further shrinking semiconductors technology and/or to provide solutions below 20 nm for processing, computing and communication devices, memory / storage devices and for further system integration;

- micro-nano-technologies as a concept for further integration and miniaturisation of sensing and actuating and to bring intelligence and networking in products and in their interfacing with the surroundings;

- nano-science and nano-technology as the ability to manipulate at molecular level to create new structures , materials, processes;

- nano-bio-technology and nano-chemistry

These lines of activities are building upon long term research on generic nano-technologies with generic application potential and on research for equipment, instrumentation and production techniques.

No common approach would fit the requirements and needs in all areas. Although one could argue for one major nano-technology programme, maturity, actors, objectives aimed at and priorities are significantly different, which in my view justifies different programmes / cluster of activities.

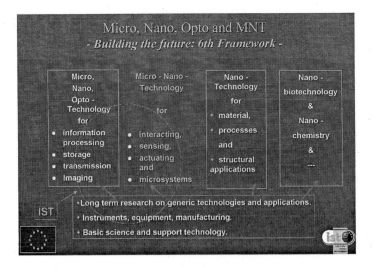

Fig. 1. Wide Range of MNT

1.2 The Sixth Framework Programme

First calls for the sixth Framework programme could be launched early in 2003. According to the Co-Decision procedure the Council and the European Parliament in the first reading adopted on December 10th, 2001 a common position on the proposal for the sixth Framework Programme of the European Commission (EC).
(see http://europa.eu.int/eurlex/en/com/dat/2001/en_501PC0500.html)

The Common Position recognises the importance of new instruments (Integrated Projects and Networks of Excellence) as an overall priority means to attain the objectives of critical mass, management simplification and European added value, and the integration of research capacities. Together with these new instruments specific targeted projects, co-ordination and other actions are continued. These modalities are not described in this article.

Networks of Excellence

The purpose of Networks of Excellence (NoE) is to strengthen and develop Community scientific and technological excellence by means of the integration, at European level, of research capacities currently existing or emerging at both national and regional level. Each Network should also aim at advancing knowledge in a particular area by assembling a critical mass of expertise. They aim to foster co-operation between capacities of excellence in universities, research centres, enterprises, including SMEs, and science and technology organisations. The activities concerned will be generally targeted towards long-term, multidisciplinary objectives, rather than predefined results in terms of products, processes or services.

Integrated Projects

Integrated Projects are designed to give increased impetus to the Community's competitiveness or to address major societal needs by mobilising a critical mass of research and technological development resources and competencies. Each Integrated Project should be assigned clearly defined scientific and technological objectives and should be directed at obtaining specific results applicable in terms of, for instance, products, processes or services. Under these objectives they may also include more long-term or "risky" research.

The activities carried out as part of an Integrated Project include research and, as appropriate, technological development and/or demonstration activities, activities for the management and use of knowledge in order to promote innovation, and any other type of activity directly related to the objectives of the Integrated Project.

Participation of the Community in Programmes Undertaken by Several Member States (Article 169)

Pursuant to Article 169 of the Treaty this instrument could cover participation in programmes implemented by governments, national or regional authorities or research organisations. The joint implementation of these programmes will entail recourse to a specific implementation structure.

2   Micro-nano-technology in Automotive Applications Examples

2.1   The Requirements

Customers are pushing to better economy of operations over the average lifecycle of the vehicle including purchase cost, fuel, road taxes, more comfort and higher vehicle performance, increased quality, reliability, comfort and access to mobile information systems. At the same time new regulations for environment friendlier transport systems are mandatory and safety standards have to be raised.

The production development cycle for automotive electronics is in most case nearly 7 years due to very high quality standards. The customer is expecting automotive electronics components to have in excess a 95% quality factor for minimum 250.000 km. Also most of the electronic components need to conform to harsh environment. The underhood temperature for components had been raised from 105 °C to 125°C and even higher for certain locations. Also the airflow used for cooling purpose had been decreased in the past.

Due to the high complexity of the new automotive systems enhanced vehicle diagnostics are necessary for an efficient and cost effective servicing. The production volumes for automotive components are only a few million pieces, which is low in relation to other e.g. telecom application, but still very high related to aeronautics applications.

2.2   Constraints and Driving applications

Safety, powertrain management, infotainment and Chassis/body control could provide an important opportunity for MNT development with subsequent exploitation in the 2008-2010 timeframe.

Ultra low emission vehicles need more and more complex Engine Management Systems. The close-loop control cycle for the different sensors should be lower than 15 msec, if the next injection cycle has to be adapted. Already today 5 or more injection strokes, which are not controlled and only set during calibration, are shaping the combustion profile. The sensors and Microsystems are performing more data processing and management. For achieving the fast close-loop control the Microsystems are required to be as close as possible to the combustion chamber. They are resisting to the harshest environment including EMC, temperature, dust, temperature cycling, vibration, etc, but still are conforming to the highest standards of reliability. New packaging methods are necessary to meet these requirements.

From the system side the flash memory of the micro-controller including the data transfer speed needs to be exponentially increased. This complexity needs new approaches for testing, validation and calibration.

For the industry the most important key is to select the right approach and new technologies and adapt to them very fast. An adaptive engine control / self-calibrating engine with high temperature and ultra fast pressure sensors and gas sensors are subjects of interest.

These applications are pushing the technologies for optical devices, for distributed multi-sensor and actuator platforms, for low cost sensor-system packaging and for sensor systems for harsh environmental conditions. In the area of passive Safety advanced Restraint System with a closed-loop control of airbag deployment system could include 3D Pre-crash detection, 3D Occupant detection, Micro inertial navigation system. Active safety is pushing for new car radar systems including merging of signals from different sources.

For an advanced Chassis / Body control a system for vehicle stabilisation could be developed including: Stability control systems encompassing low-g accelerometers and angular rate sensors, advanced tire pressure measurement system (TPMS) encompassing several sensors to monitor vibration, vehicle load etc., advanced braking system. System integration technologies are driven by cost effective systems including packaging, signal conditioning and interfaces aspects.

2.3   Other aspects

Integration of future plurality of automotive electronic systems into a few larger systems requires standardisation of interfaces and communication methods and protocols. Also, the legislative situation in Europe with respect to safety issues for tire pressure measurement systems, collision waning and avoidance systems is expected to be addressed in future activities.

3   Summary

MNT is driving a growing set of developments. Some activities may profit from European collaboration in the domain or related domains. The Framework 6 programme will offer opportunities for new activities. Within the area we are counting on your inputs for building a common vision and preparing the future.

Background Information

Thomas Reibe is working at the Directorate-General Information Society of the European Commission. The views expressed in this article are those of the author and do not necessarily reflect the official European Commission's view on the subject.

References

[1] Commission proposals Document 501PC0500, http://europa.eu.int/eur-lex/en/com/dat/2001/en_501PC0500.html
[2] Commission communication (21 January 2002) on "Integrated Projects and "Network of Excellence", http://europa.eu.int/comm/research/nfp/networks-ip.html

1.4 Regulations for Automotive Applications

T. Goernig[1], S. Krueger[2]

[1]Conti Temic microelectronic GmbH
Ringlerstrasse 17, 85057 Ingolstadt, Germany
Phone: +49/841/881-2493, Fax: +49/841/881-2420
E-mail: thomas.goernig@temic.com

[2]VDI/VDE-IT
Rheinstrasse 10B, 14513 Teltow, Germany
Phone: +49/3328/435-221, Fax: +49/3328/435-256
E-mail: krueger@vdivde-it.de

Keywords: regulation, safety systems

Abstract

Regulations are helping to standardise specific questions on automotive technologies and applications, helping to reach objectives, use scale effects and assuring specific reliability. It is difficult to get an overview on existing regulations and by offering the following lists we would like to help to broaden common sense.

We are planning to expand this section into other reglemented fields relevant for microsystems, therefore looking for support on information for e.g. powertrain regulations.

Regulations

The following table is for information only and gives an overview about major regulations that are effective for road vehicles and passenger safety systems. The table has been compiled to the best knowledge and does not claim to be complete. Anyhow the authors can not take any liability for the information given herein. The table shall be revised periodically as a guide for the interested reader, in case of additions and/or corrections comments are welcome by the authors.

Country	Regulation	Requirement
Germany	StVZO 30 StVZO 35 a	General requirements for occupant safety 3-point belts for limousines, 2-point belts f or convertibles

1.4 Regulations for Automotive Applications

Country	Regulation	Requirement
Europe	ECE-R 11	Door locks and door retention components.
	ECE-R 12	Impact protection for the driver from the steering control system.
	ECE-R 14 EWG 76/115	Seat belt assembly anchorages
	ECE-R 16 EWG 77/541	Seat belt assembly.
	ECE-R 17	Seating systems and anchorages
	ECE-R 21	Interior trim.
	ECE-R 25	Head restraints.
	ECE-R 26	External components
	ECE-R 32	Rear impact protection.
	ECE-R 33	Front impact protection.
	ECE-R 34	Fire prevention (Fuel system)
	ECE-R 42	Bumpers.
	ECE-R 44	Child restraint systems.
	ECE-R 66	Bus body joint strength.
	ECE-R 73	Side protection for trucks and trailers
	ECE-R 80	Seat stiffness of busses
	ECE-R 93	Front under ride protection.
	ECE-R 94	Occupant protection at front impact
	ECE-R 95	Occupant protection at side impact
	ECE-R 107	Double deck busses, general requirements
	III/5021/96 EN COM(2001)389	Protection of pedestrians and cyclists
USA/CDN	FMVSS101-135	General requirements.
	FMVSS201	Occupant protection in interior impact.
	FMVSS202	Head restraints.
	FMVSS203	Impact protection for the driver from the steering control system.
	FMVSS204	Steering control rearward displacement.
	FMVSS205	Glazing materials.
	FMVSS206	Door locks and door retention components.
	FMVSS207	Seating systems.
	FMVSS208	Occupant crash protection
	FMVSS209	Seat belt assemblies.
Japan	31	

Country	Regulation	Requirement
Switzerland	23 / 3	
	FMVSS210	Seat belt assembly anchorages.
Japan	37	
	FMVSS212	Windshield mounting.
	FMVSS213	Child restraint systems.
	FMVSS214	Side impact protection.
	FMVSS216	Roof crush resistance.
	FMVSS217	Bus emergency exits and window retention and release.
	FMVSS218	Motorcycle helmets.
	FMVSS219	Windshield zone intrusion.
	FMVSS220	School bus rollover protection.
	FMVSS221	School bus body joint strength.
	FMVSS222	School bus passenger seating and crash protection.
	FMVSS223	Rear impact guards.
	FMVSS224	Rear impact protection.
	FMVSS225	Child restraint anchorage systems.
	FMVSS301	Fuel system integrity.
	FMVSS302	Flammability of interior materials.
	FMVSS303	Fuel system integrity of compressed natural gas vehicles.
	FMVSS304	Compressed natural gas fuel container integrity.
	FMVSS305	Electric-powered vehicles: electrolyte spillage and electrical shock.
	FMVSS500	Low-speed vehicles.
Australia	ADR 4 d	Seat belt strength.
	ADR 5 b	Position and static strength of seat belt anchorages.
	ADR 34 a	Position, static strength and accessibility of child restraint systems anchorages.
Japan	22.3	3-point belts for front seats; strength.
Sweden	109	3-point belts for outboard seats, 2-point belts for middle seat, otherwise compatible with ECE-R 14 and ECE-R 16.
Switzerland	23/3	3-point belts for front seats, otherwise compatible with ECE-R 14 and ECE-R 16.
Saudi Arabia	SAS 297/82	Warning system for unbelted driver.

2 Microsystems Technologies

2.1 Microsystems Technology – a New Era Leading to Unequalled Potentials for Automotive Applications

F. Solzbacher

First Sensor Technology GmbH
Carl-Scheele-Strasse 16, 12489 Berlin, Germany
E-mail: solzbacher@first-sensor.com

Abstract

This article gives an introduction to the topics covered in the AMAA 2002 almanac. General concepts and terms are introduced in order to give an overview over the system and main components and issues of microsystems technology. Ranging from materials and micromachining technologies over components like sensors, package and signal processing to complex micro mechatronics and reliability issues all key topics are introduced.

Microsystems Technology (MST) – a Complex Game of Materials, Technologies and Design

Microsystems is a term evolving from the former microperipheric technology and first introduced in Europe the late 1980's and early 1990's. Fine mechanics and microelectronics were both approaching each other, in order to merge into a new generation of miniaturized micro systems consisting of sensing, processing and actuation components. Europe (MST), the USA (MEMS-MicroElectroMechanicalSystems) and Japan (MicroMachines) are the key drivers in the development and market introduction of MST products.

Basically, the general concept is to use new materials, mostly well suited for low cost mass production purposes and machining technologies, to form three basic building blocks of every micro system:

- Sensing / actuation element
- Package
- Signal Processing

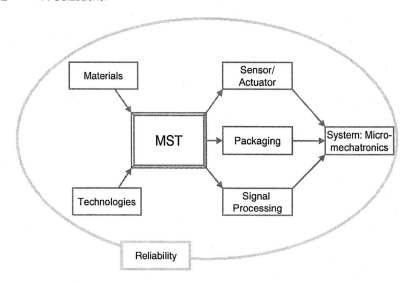

Fig. 1. Microsystems Technology: Key Components and Issues

Some key challenges faced by this new generation of products are:

- Price – new technologies usually have to match or outperform existing technologies at 10-20% lower price
- Media compatibility – most of the promising applications today require high media compatibility, since the processes to be controlled frequently envolve aggressive liquid or gaseous media
- High temperature compatibility – most of the key applications in high volume markets are found in automotive control systems with high temperature requirements
- Integration of multiple functionalities – apart from the general cost reduction a cheap microsystem brings about compared to conventional solutions, frequently logistics and quality control can be simplified when integrating several components in one system: e.g. multiple sensor arrays
- Reliability – failure modes and reliability of microsystems are already well investigated in many areas

MST products have entered the market in far more depth than commonly known: a huge variety of MST-components have become standards, such as acceleration sensors for airbag ignition, pressure sensors for a variety of automotive and medical applications (e.g. blood pressure sensors), inkjet printheads in office printers and many more. The new 7series BMW marks a new milestone in the use of MST-components with a total of 70

2.1 Microsystems Technology

devices installed in the car. This is but the start of a new generation of products.

Materials

Advances in materials science and processing are the corner stone of each MST product. Three main groups of materials are to be distinguished: materials for the package, the actual device, the electronics and the mechanical/electrical connection between these.

Progress in semiconductor processing has evolved in a number of substrate materials, pre-destined for the use in micro structured devices, such as Silicon, Silicon-on-Insulator (SOI), Silicon Carbide and Gallium Arsenide.

- Si: Silicon is today's most used semiconductor material. Substrate and processing technology and research has reached a level of perfection unmatched by any other material. Inherent material limitations are frequently overcome by continued optimisation.
- SiC: SiliconCarbide is one of the most promising materials for ultra high media resistance and high temperature compatibility
- SOI: Silicon on Insulator material is today's most advanced Silicon based substrate technology : using an integrated insulation layer it combines all the advantages of the most well established Silicon-technology with high media and temperature compatibility
- GaAs: GalliumArsenide being a so called wide-bandgap semiconductor is well suited for high temperature applications, but has little developed micro machining processes at hand.

Pricing and reliability considerations have lead to an almost exclusive use of Si-based micro machined devices. Packaging and assembly has focused on ceramics, PCB-technology, MCM's

Bulk vs. Surface Micromachining

A key process for MST is the 3D-machining technique of semiconductor materials leading to the miniaturized structures constituting the sensing, actuating or other functional parts.

The two main processes to be stated here are bulk and surface micro machining. In bulk micro machining an wet etching agent such as KOH, EDP or TMAH or alternatively, a dry etching process using e.g. SF_6 gas is used to etch 3D structures directly into the bulk Si-wafer. Structures of up to wafer thickness can easily be realised. Surface micro machining uses sandwiches of so called sacrificial and functional layers on top of the Si-wafer. Removal of the sacrificial layers results in free standing structures with thickness of about 1-5 µm.

Sensors/Actuators

Sensors and Actuators are the core of the complete micro system: usually a naked chip manufactured in bulk or surface micromachining used for the detection of a physical or chemical quantity or some actuation principle, like the dosage of ink droplets in inkjet printheads.

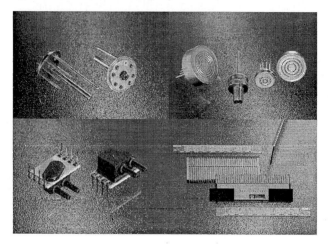

Fig. 2. Examples of MST-based Sensors and Actuators in Various Packages

2.2 Fabrication Technologies for 3D-microsensor Structures

F. Solzbacher

First Sensor Technology GmbH
Carl-Scheele-Strasse 16, 12489 Berlin, Germany
E-mail: solzbacher@first-sensor.com

Abstract

This article summarizes some of the main processing technologies for the three dimensional structuring of silicon based MEMS devices. It is intended as a brief overview over the most important technologies such as bulk and surface micromachining. Furthermore, a selection of key characteristics and mechanisms to be taken into consideration when designing and fabricating MEMS devices.

Introduction

Whereas integrated circuits only require planar structures of a few nanometers to micrometers, micromechanical devices usually exhibit thicknesses of several micrometers to millimetres. The general approach towards the processing of the devices however remains largely unchanged to most other semiconductor based technologies: functional layers and structure are created by the succeeding deposition/implantation and lithographic structuring of layers and bulk material.

Bulk Micromachining

Most micromechanical devices require 3D machining of the bulk silicon material with etching depths of up to wafer thickness. Generally, three basic etching process types can be distinguished:

- Isotropic etching
- Anisotropic etching
- Electrochemical etching

Isotropic Etching

Isotropic etching processes etch the bulk silicon material equally in all crystal directions. Isotropic etching requires a sequence of an oxidation process of the silicon followed by a removal of the oxide layer. $HF/HNO_3/H_2O$ is amongst the most frequently used wet etchants. Masking layers can be e.g. Si_3N_4 or Au/Ti layers. The nature of the isotropic etching process results in underetching of the masking layers with extensions about equivalent to the etching depth.

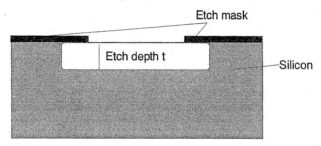

Fig. 1. Underetching of Etch Mask During Isotropic Etching

Alternatively, dry etching processes can be used, which combine a gaseous etchant with ion bombardment thereby physically removing material. These so called RIE (Reactive Ion Etching) – processes use e.g. SF_6/O_2 as etching agents. Recent years have also shown the upcoming of a new generation of Deep Reactive Ion Etching processes and equipment (DRIE) which allows the etching of deep trenches with high aspect ratio with low underetching. The "Bosch process" is amongst the most commonly used process for DRIE.

Anisotropic Etching

Anisotropic etching process exhibit a strongly varying etch rate with crystal orientation of single crystal semiconductor material. They are bases for the majority of micromechanical devices and structures. Exclusively basic etchants feature this characteristic: examples are potassium hydroxide (KOH), sodium hydroxide (NaOH), Ammoniumhydroxide (NH_4OH/TMAH) and organic solutions such as Ethylendiamin-Pyrocatechol (EDP). Today, for safety and health reasons almost exclusively KOH is used for this purpose.

Temperature and concentration of the KOH solution as well as the doping concentration of the silicon material are crucial factors in the determination of the etching rate and anisotropic characteristics.

Crystal Orientation of Etch Rate

Anisotropic etching agents feature a two orders of magnitude smaller etch rate in (111) direction of the crystal compared to the (100) and (110) direction. As a rough model, the number of free dangling bonds in the respective crystal directions influence the bonding energy of Silicon atoms thereby determining the etch rate. Thus, in (100) and (110) material structures limited by (111) surfaces can be created. These surfaces are oriented to each other at characteristic angle of 54,7° (100-material) or 90° (110-material).

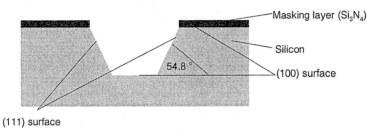

Fig. 2. Anisotropic Etching Using KOH in (100) Silicon Material

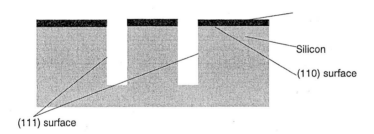

Fig. 3. Anisotropic Etching Using KOH in (110) Silicon Material

Passivation Layers

Passivation layers used in KOH etching are almost exclusively Si_3N_4 or SiO_2 layers. The higher etch rate of SiO_2 in KOH leads to the exclusive use of Si_3N_4 for deeper etching structures. These layers are deposited via Low-Pressure-Chemical-Vapor-Deposition (LPCVD) or Plasma-Enhanced-CVD (PECVD) with LPCVD layers being the more dense and KOH resistant.

Convex-corners

Convex corners in micromechanical structures expose crystal surfaces with high etch rate to the KOH etchant resulting in high underetching of the masking layer. Underetching in this case depends strongly on etchant, etch time and depth as well as the surface characteristics of the silicon wafer.

In order to prevent this effect from destroying the desired geometrical shape, additives like Isopropylalcohol (IPA) can be added to the KOH changing the etchant composition and etching mechanism. Usually, however, compensation structures are placed on the convex corners which are succeedingly consumed during etching., thereby protecting the original geometry of the structure.

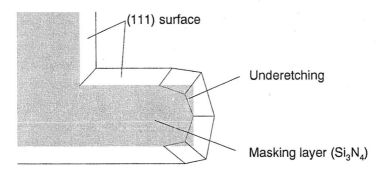

Fig. 4. Convex Corner Underetching

Electrochemical Etching

Electrochemical etching uses the Silicon wafer as anode in an HF-Electrolyte. At sufficiently high currenty, oxidation of the silicon occurs which is dissolved by the HF-solution. Lowly doped silicon material is not exhibiting a notable etch rate. Thus, e.g. at n+n and n+p structures etching stops at the n-layer.

In the production of Silicon based pressure sensors this process is frequently used by using wafers with a thin low-n-doped epitaxial layer determining an etch stop depth and thus the thickness of the sensor membrane.

This process can also be carried out in anisotropic solutions like KOH/H_2O. In this case the pn-junction resulting from the epitaxial layer is put into reverse bias voltage with the complete potential difference occurring across the pn-junction. Until the KOH-etching process reaches the junction, it runs just like a normal KOH-etching process. At the junction the potential changes resulting in a passivation layer stopping the etching process.

Surface Micromachining

In surface micromachining 3D-structures of only a few micrometers thickness are created by succeedingly depositing and etching sandwiches of active layers and separating sacrificial layers. At the end of the process run, these sacrificial layers are removed leaving the desired 3D structures on the wafer surface.

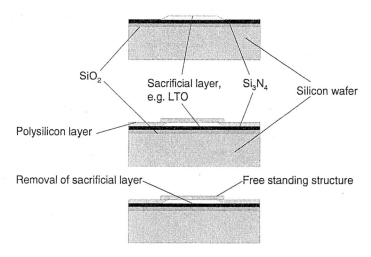

Fig. 5. Principle Schematic of Surface Micromachined Structures

Conclusion

The choice of 3D-micromachining technology depends strongly on the application field and design of the MEMS product to be manufactured. Cost considerations usually strongly favour the cheap KOH wet etching process allowing batch processing of large quantities of wafers. Nevertheless, surface micromachined pressure sensors or gyroscopes as well as a number of acceleration sensors have achieved good commercial success.

The decision on which fabrication technology to use however still has to be made individually for each application and design case.

2.3 Signal Processing for Automotive Applications

T. Riepl, S. Bolz, G. Lugert

Siemens VDO Automotive AG, SV P ED Innovation Centre
Osterhofener Strasse 14, 93055 Regensburg, Germany
E-mail: thomas.riepl@at.siemens.de

1 Sensor Signals as Input for Automotive Control Units

Engine and Transmission Control, as well as many other functions of automotive systems (safety, comfort, environmental protection, ...) require effective control concepts in order to fulfill the requirements from legislation and the customers. These control concepts can be basically reduced to a control loop: sensor measures the controlled variable - control unit processes the sensor input and drives the actuator - actuator adjusts the controlled variable.

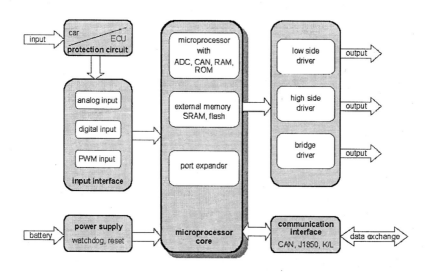

Fig. 1. Block Diagram of a Typical Engine Control Unit (ECU) [1]

In modern cars up to 70 control units are working. Most of the implemented control concepts are rather complex and involve a number of sensor signals to handle. As an example the block diagram of an engine control unit (ECU) is shown in Fig. 1. A typical ECU provides several interfaces to sensors:

- analog input
- digital input (on/off)
- pulse width modulated (PWM) input
- CAN interface

In order to protect the electronics from overvoltage all sensor signals have to pass input protection circuits (passive: R, RC circuits; active: specific semiconductors) and filters. These circuits suspend noise and limit the signals to the allowed input range of the microcontroller (0 V to 5 V). After analog to digital conversion, the sensor signals are typically processed fully digital by the microcontroller.

In addition to these wire-based interfaces, also optical and telemetric interfaces are under development (see chapter 4). Today a typical control unit is able to handle the following signals:

Analog signals

Analog signals are the input from analog sensors like lambda probe, pressure sensor and potentiometer. The signal information is described be the voltage with a typical range of 0 V to 5 V (typically 0.5 V and 4.5 V for a defined range of the measurand). This normalisation of the input voltages requires that analog signal processing has to be included in the sensor.

Pulsed signals

A typical example for pulsed signals is the output of a inductive rotational speed sensor. Its voltage range is up to 100 V. Usually the signal is converted to a digital signal and then be handled by a counter. Typical ECUs have specific input interfaces that perform this conversion. Another example for pulsed signals is the pulse width modulated output of e.g. mass air flow sensors and manifold absolute pressure sensors.

Digital signals

The digital input interface handles the digital signals (on/off) of switches with a voltage range of 0 V to battery voltage. The voltage is limited to 5 V by the input protection circuit. In addition to this, also digitalised pulsed signals like that of a hall probe with digital output are handled by the digital input interface. The voltage of digitalised pulsed signals ranges usually from 0 V to 5 V.

Data transferred via bus

The communication interface of the ECU is designed to exchange digital information with sensors via bus. This requires a smart bus capable sensor with integrated analog to digital conversion and communication interface. The signal form has to follow the bus protocol, usually CAN.

2 Signal Processing

The full sequence from primary signal to digital information is explained in brief in this chapter. Also the so called smart sensors are introduced. In addition, the advantages and disadvantages of monolithic integration, as well as the trend towards systems integration are discussed.

2.1 From Primary Electrical Signal to Digital Information

The basic sequence of signal processing is illustrated in Fig. 2. The measurand is transformed to a primary signal by the sensor element. In the analog signal processing step, the signal is filtered, amplified and normalised - usually to a voltage range of 0 V to 5 V. After analog to digital conversion, the signal can be processed fully digital by the microprocessor of the ECU.

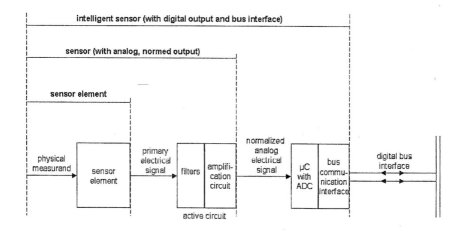

Fig. 2   Basic Sequence of Signal Processing and Definition of Terms: Sensor Element - Sensor - Intelligent/Smart Sensor [2]

Besides these basic steps, signal processing includes also linearisation and compensation of perturbances, which can be performed in an analog or digital way. Usually the analog signal processing unit includes also a calibration circuit. In some cases it is advantageous to perform calibration digitally (software calibration).

Sensor signals are affected by non-linearities and interfering quantities, like temperature. In general, the achievable characteristics of a sensor depend on the structure and complexity of the individual device. Mostly the following structures are used for analog compensation [6]:

- serial structure (e.g. temperature compensation by a resistor network)
- parallel structure (e.g. differential structure)
- loop structure (e.g. closed-loop accelerometer)

The most common is the serial structure, while the loop structure is only used when it is impossible to achieve the requirements by other means. Detailed information on serial structure and loop structure can be taken from the examples, given in chapter 4.

Depending on application, the signal processing is performed partially or fully by the sensor. Corresponding to this, the ECU receives either a partly processed analog/digital signal, that requires post-processing, or it receives fully processed digital information via bus. In the following the basic steps are explained in more detail:

Primary electrical signal

The primary electrical signal is described by voltage/current (analog signals), or by frequency (pulsed signals). It is created by the sensor element and has a direct, in some cases linear correlation to the physical or chemical measurand.

Amplification

As the primary analog signals are typically rather weak, amplification is required. Mostly amplification is accompanied by offset correction and normalisation of the signal to a range between 0 V and 5 V.

Filtering

Usually analog signals come along with high frequency noise. Therefore a low pass filter is often part of the analog signal conditioning. The cutoff frequency of the filter has to respect the required dynamics of the sensor as well as the sampling frequency of the AD converter (anti aliasing filter).

Compensation

Most physical and chemical effects that are used to create the primary electrical signal are affected by perturbing effects like temperature. Especially in automotive applications, where temperature varies over a wide range, compensation of these undesired effects is required. This can be done in different ways:

- analog compensation by using the perturbance effect itself as a reference (e.g. compensation of the temperature dependence of a piezoelectric sensor element)
- analog compensation by an adapted (trimmed) compensation circuit
- digital compensation via look up table using the information of an additional temperature sensor e.g. via a diode as a reference

Calibration

For high accuracy measurements, calibration is mandatory in order to compensate the manufacturing tolerances of sensor element and analog signal processing circuit. This can be done on hard ware level, e.g. by laser trimming but also on software level, like by fine tuning of the lookup table. During calibration also the temperature compensation can be adapted to the specimen's characteristics.

Linearisation

Usually a sensor signal is expected to have a linear correlation to the measurand. However, many physical/chemical effects that are used to create the primary signal are of nonlinear nature. Therefore most sensor signals are linearised during signal processing. Linearisation can be part of the analog signal conditioning, but is often performed in the digital stage using a lookup table.

Analog to Digital Conversion

This step requires a normalised signal with controlled offset. The analog to digital conversion can be performed by the (smart) sensor or by the microcontroller of the control unit.

2.2 Smart Sensors

According to the degree of implemented signal processing, sensing devices are usually referred to as sensor element, sensor or smart sensor. A definition of these terms has been given in [2] and is illustrated in Fig. 2. The sensor element transforms a measurand (temperature, pressure, ...) into a primary electrical signal. This analog signal is usually characterised by voltage and often depends linearly from the measurand. Otherwise, linearisation either analog or digital, is part of the signal processing. An active circuit amplifies and normalises the primary signal, typically to a range of 0 V to 5 V. In many cases also a compensation of temperature effects is included in this analog signal processing step. When we talk about a sensor, we usually think of sensor element plus active circuit for analog signal processing. If the sensing device has included on top also analog to digital (AD) conversion and a bus interface it is usually called an intelligent or smart sensor.

In general a sensor element is sensitive not only to the intended measurand but also to other effects, often being a perturbation to the primary signal. The most critical perturbing effect in automotive applications is usually temperature. In cases where the relation between output and the perturbing effect is simple compensation can be done by analog means, i.e. by incorporating a temperature sensing element in the sensor package which is used to apply a measure of correction. In cases where there are a number of perturbing effects, especially if they are nonlinear, digital techniques must be applied to obtain a "true" signal. In these cases it is often useful to implement signal processing and compensation in a smart sensor device.

Fig. 3 shows schematically the concept of a smart sensor. The signal to be measured is obtained by a sensor whose output, together with the outputs from sensors for the perturbing variables is passed via analog to digital converter to a correction unit. This is programmed with a description of the transfer functions for each variable, obtained during initial calibration. Examples for this concept are NO_x sensor and hall sensor with digital output.

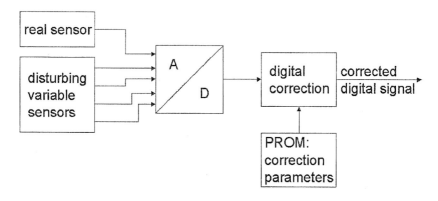

Fig. 3. Concept of a Fully Smart Sensor with Digital Compensation of All Perturbances [3]

2.3 Monolithic Integration

Classical sensor concepts were typically hybrid solutions consisting of a sensor element and a separate analog signal processing circuit, e.g. realised in surface mount technology. With the advent of micromachined sensors based on silicon substrates it is now in principal possible to integrate the signal processing onto the silicon of the sensor element. This comes along with the following advantages:

- reduction of interconnects (cost advantage, improved reliability)
- reduction of total sensor size
- signal processing circuit is manufactured as an integrated circuit (cost advantage, perfect matching of resistors, ...)

On the other hand, there are some disadvantages of monolithic integration:

- poor compatibility of process technologies for micromachining and manufacturing of the analog/digital circuit
- limited synergy between process for manufacturing the electric structures of sensor element and analog signal processing
- depending on application and design, the analog signal processing circuit might be subjected to aggressive media and/or high temperature, when integrated monolithically
- in some cases, still some external resistors are needed for calibration purposes (SW calibration can avoid this disadvantage)

Hence, it still depends on the specific application, but also on the expected production volume whether monolithic integration provides an advantage compared to hybrid solution. As a compromise, we find very often a two-chip solution, where the sensor element is on one die and the analog signal processing on the other one. With this modular approach, most of the advantages of monolithic integration can be used, while the technology related disadvantages are avoided. Furthermore, the analog signal processing for different sensor applications can be performed by one standard signal processing IC. This generates production volume, which decreases cost. Additionally the engineering effort is reduced, as basically only the sensor element itself has to be designed specifically.

Despite these advantages of two chip solutions, the overall trend heads towards monolithic integration of sensor element, analog signal processing and even analog to digital conversion followed by digital signal processing and bus communication.

2.4 Systems Integration

With the increasing complexity of vehicle systems and subsystems it becomes apparent that the output of a sensor may be required by several subsystems. It thus becomes attractive to integrate the subsystems in a bus architecture with smart sensors that communicate directly with the bus and therefore avoid sensor duplication. In early electronic engine management systems even the ignition and fuelling systems were completely separate, and had separate sets of sensors. Today manifold pressure sensor, engine speed and position sensors are used in the ECU to determine fuelling, ignition, and exhaust gas recirculation requirements. Similarly it is obvious to integrate the sensing requirements of braking, steering and suspension, because of the interactions of braking and cornering forces with suspension levelling. Pursuing this line of development, one can think of a total vehicle control system combining driver's input with various automatic safety controls and guidance systems for a complete control of the vehicle dynamic. The system shown in Fig. 4 was already proposed in 1993 (see [4]). In the recent years the major aspects of this proposal have been implemented.

The general trend towards systems integration of sensors is a main driver for the market penetration of smart sensors.

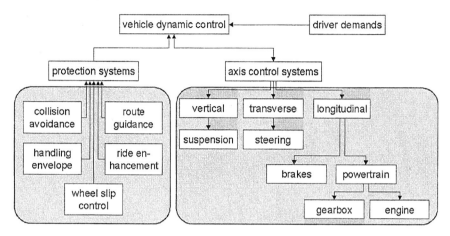

Fig. 4. Hierarchical Vehicle Control System [4]

3 Examples for Analog Signal Processing

3.1 Linear Oxygen Sensor

The precise, fast and wide range measurement of exhaust gas oxygen concentration is a key element to the development of low emission combustion engines. The signal processing for a linear oxygen sensor follows mainly a closed loop concept and is shown in Fig. 5.

The basic sensing element of a linear oxygen sensor is a Nernst Cell. One electrode is exposed to the measured gas, the other to an oxygen reference. The sensing element is confined to a measurement cavity, which is located inside the sensor and linked to the exhaust gas by a gas diffusion barrier. The sensor is fitted with another pair of electrodes (pump cell), which allows to transport oxygen ions from the exhaust gas into the cavity and vice versa. Additionally it contains a heating element to maintain a constant operating temperature.

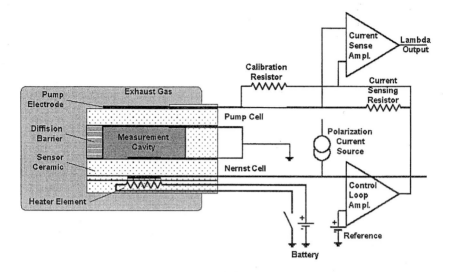

Fig. 5. Linear Oxygen Sensor: Basic Construction and Closed Loop Analog Signal Processing

In the close vicinity around Lambda = 1 the Nernst Cell transfer function is linear. If the oxygen concentration in the cavity is maintained at Lambda = 1, the Nernst Cell will report minute deviations. Comparing its output voltage with a reference gives the necessary error signal to establish a pump current control loop. The pump current used to transport oxygen ions into and out of the measurement cavity correlates to the amount of oxygen flowing through the diffusion barrier. This in turn is related to the difference of oxygen concentration between exhaust gas and cavity. As the lambda value inside the cavity is maintained at 1 it is now possible to derive the oxygen concentration of the exhaust gas from the pump current. Tolerances of the diffusion barrier are measured during manufacturing. The sensor is then fitted with an individual calibration resistor. Reading the resistor value allows either a gain adjustment of the current measurement or a software correction of the measured current value and by this a compensation of the manufacturing tolerance.

Due to its construction the diffusion barrier resistance changes with temperature. The temperature – impedance characteristics of the sensor equals an NTC, making it ideal for detecting the sensor temperature. A temperature stabilisation of the sensor can now be established by reading the sensor impedance, comparing it against a target value, and controlling the heater power.

Having applied all these measures, the pump current is proportional to the oxygen partial pressure of the exhaust gas. The precision of its measurement defines the sensor quality. Several actions have to be taken to improve accuracy:

- suppression of common mode signal
- compensation of sensor manufacturing tolerance
- low pass filtering to remove undesired high frequency components from the output signal

Analog signal conditioning is followed by AD-conversion. Communication of the sensor signal to the micro controller is performed e.g. by Serial Peripheral Interface (SPI). A more detailed description of sensor, analog signal processing and also the digital interface to the microcontroller can be taken from [5].

3.2 Piezoresistive Pressure Sensors

One of the first micromechanic sensors to enter the automotive market was the piezoresistive sensor for measuring manifold absolute pressure (MAP). The sensor element itself is basically a piezoresistive film on a silicon membrane. The piezoresistive film forms a Wheatstone bridge with the bridge voltage as a primary electrical signal. Membrane and resistors are designed to achieve a voltage response which is linear to the pressure that is applied to the membrane. In general the analog signal processing involves components for amplification, offset correction, and compensation of temperature effects on the offset and sensitivity of the sensor.

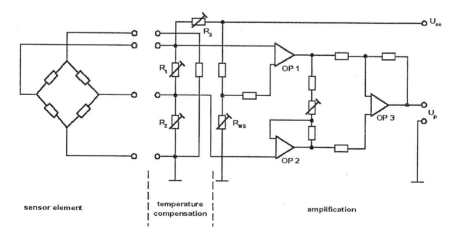

Fig. 6. Analog Signal Processing Unit for a Piezoresistive Pressure Sensor [6]

Fig. 6 shows the analog signal processing unit for the pressure sensor. The sensor element is followed by a resistor network which affects the temperature compensation. Ideally, the correction of the temperature effects on offset and sensitivity should be accomplished independently of each other, but in practice this is very difficult to achieve and depends strongly on the electric circuit design. Here the offset temperature coefficient is corrected by trimming the resistors R_1 and R_2, while temperature influences to the pressure sensitivity of the sensor are compensated by the temperature independent resistor R_3. The necessary values of the resistors are calculated for each individual sensor according to the calibration data at different temperatures and pressures. Finally, the temperature corrected sensor signal is amplified by a symmetrical amplifier stage with an amplification factor of up to 50.

4 Signal Transfer

Which kind of sensor should be used depends among others on the necessary features for signal transmission and on the kind of signal processing applied. The major signal forms can be described as follows:

- amplitude analog
- frequency analog (modulation of frequency or pulse width)
- direct digital

The main characteristics of these signal forms are listed in Table 1.

2.3 Signal Processing for Automotive Applications

feature	amplitude analog	frequency analog	direct digital
possible static accuracy	restricted	unlimited	unlimited
dynamic characteristics	in general very good	limited by conversion speed	limited by net data transfer rate
noise immunity on signal transmission	small	good at FM transmission	good; option for digital error correction
possible arithmetic operations	limited or expensive	division/integration easy feasible	any operation with arithmetic processors
connection to digital control units	by analog to digital conversion	e.g. by frequency counter	direct at suitable code and signal level
error correction	improvement only by analog averaging	on frequency counting natural redundancy	with error-correcting codes
galvanic signal separation	very expensive	easy	easy

Table 1. Comparison of the Features of Signal Forms [6]

A rather novel development, but with high potential to reduce wiring harness is the so called DC BUS. The DC BUS high-speed multiplex technology transfers digital information inside the car using a single wire which integrates the functionality of 14 V or 42 V power net and communication. The use of a single wire for battery power and for communication allows to reduce the amount of wiring and will also enable to add optional modules after assembly of the car [7].

Up to now mainly copper cables have been used for the transfer of safety relevant information within cars. On the other hand optical networks with a ring topology for media oriented data transfer like D2B and MOST (Media Oriented Systems Transport) are already in production since 1997. The byteflight system, which is ready to launch will be the first network with an optical physical layer for passive safety, comfort and body control [8]. Also telemetric communication is under investigation. The main obstacle for the introduction of telemetric data transfer are EMC related safety issues. Hence, this technology is up to now mainly discussed for the so called intelligent tire [9].

5   Outlook/Trends

In general smart sensors are likely to be used increasingly, because of their ability to compensate for errors and changes in measurement conditions. Their digital output renders them also capable of communicating directly with a bus, which opens up possibilities of system integration and the sharing of sensor data between subsystems.

This trend is assisted by a general trend towards a decentralised E/E architecture with distributed intelligence within almost all sensors and actuators. Blocking point for this trend up to now is the cost of distributed intelligence, especially if the potential to reduce wiring harness is not utilised. Regarding this issue, the market entry of the DC BUS as well as the optical networks will probably accelerate the trend towards smart sensors. With increasing complexity of the sensors, also integrated diagnostics will be a requirement of the future.

As a long-term trend, the additional optical and telemetric communication channels will also affect the data processing inside sensors. For using these channels, additional conversion stages have to be included in the sensors. Especially telemetric signal transfer seems attractive as it enables to create self sustaining sensor systems without any cables [10] which have a high flexibility for mounting locations.

References

[1] S. Bolz, G. Lugert "Aufbau moderner Steuergeraete" in Elektronik im Kraftfahrzeug, Walliser u.a., expert Verlag Renningen(2000).

[2] Sensortechnik, H.R. Traenkler, E. Obermeier (Hrsg.), Springer Berlin (1998).

[3] P. Cockshott "Automotive Sensors" in Sensors A Comprehensive Survey **8**, edited by W. Goepel, J. Hesse, J.N. Zemel, VCH Weinheim (1995).

[4] "Electronic System Architecture", Automotive Engineering, April 1993, p. 15-17.

[5] S. Bolz "A Novel Interface for Linear Oxygen Sensors" in ATTCE 2001 Proceedings (2001).

[6] H.R. Traenkler "Signal Processing" in Sensors A Comprehensive Survey 1, edited by W. Goepel, J. Hesse, J.N. Zemel, VCH Weinheim (1989).

[7] Y. Maryanka "14V/42V Power Line Communication for Automotive" in Advanced Microsystems for Automotive Applications 2001, edited by S. Krueger, W. Gessner, Springer Berlin (2001).

[8] J. Wittl, K. Panzer, H. Hurt, E. Baur "Transceivers for Optical Networks in Automotive Applications" in Advanced Microsystems for Automotive Applications 2001, edited by S. Krueger, W. Gessner, Springer Berlin (2001).

[9] M. Loehndorf "Wireless Tire Sensors Based on Amorphous Magneto-Elastic Materials" in Advanced Microsystems for Automotive Applications 2001, edited by S. Krueger, W. Gessner, Springer Berlin (2001).

[10] G. Hettich, V. Vieweger, J. Mrowka, G. Naumann "Self Supporting Power Supply for Vehicle Sensors" in Electronic Systems for Vehicles, VDI Verlag Duesseldorf (2001).

2.4 Materials for Electronic Systems in Automotive Applications – Silicon On Insulator (SOI)

P. Seegebrecht

Christian–Albrechts-Universitaet zu Kiel, Lehrstuhl für Halbleitertechnik
Kaiserstrasse 2, 24143 Kiel, Germany
Phone: +49/431/8806075, Fax: +49/431/8806077
E-mail: ps@tf.uni-kiel.de, URL: www.tf.uni-kiel.de/~ps

Abstract

Much of the electronics of future cars will be physically integrated together with sensors and actuators to mechatronic systems. Depending on the location of such a system, the maximum temperature experienced by the electronics may exceed the mil-specified limit of 125°C. The strict low cost requirements in the automotive industry favour only silicon-based electronic systems as promising approach for mass production. In this article, first of all the influence of temperature on device characteristic will be discussed. Then, the advantage of SOI devices with respect to elevated operation temperatures will be demonstrated. Finally, processes for the fabrication of SOI materials are presented.

1 Introduction

Electronic systems play an increasingly important role in modern cars. The average electronics value in the car has risen from 7% in 1990 to 17% in 2000 and is bound to reach 25% in the near future [1]. This trend is driven by requirements concerning pollution control, safety, comfort etc.. Much of the electronics of future cars will be physically integrated together with sensors and actuators to mechatronic systems. The modular architecture of such systems offers cost reduction (fewer components, batch production), reduction of volume and weight, new functionality (self test, accuracy test), and higher reliability (fewer plugs and cables, fewer components). Examples of this kind of developments are engine management, transmission control, emission control, or ABS systems.

Direct mounting of electronics to the aggregates means, that the electronic components have to withstand high ambient temperatures and rapid temperature cycles between –40°C and a maximum temperature, which in several cases exceeds the mil-specified limit of 125°C. Taking into account that power losses inside the electronics lead to self heating of the devices, Table 1 gives examples for the junction temperature of electronic components at various locations within the car.

Location	Temperature Range (°C)
Cabin, chassis	-40 85
Engine compartment	-40 150
On the engine	-40 200
Wheel mounted	-40 300
Combustion chamber	-40 500
Exhaust system	-40 800

Table 1. Temperature Conditions for Distributed Electronics [2]

Performance and lifetime of semiconductor devices are very sensitive to these upper temperature limits. In the following, the influence of temperature on material parameters and device characteristics will be described with respect to functional limits.

2 Influence of Temperature on Device Characteristics

There are several physical parameters, which need to be considered in order to determine the suitability of a semiconductor material for high temperature operation. The most important ones are bandgap and thermal conductivity. The intrinsic carrier density n_i, i.e. the electron and hole density of an undoped semiconductor, depends exponentially on the temperature and bandgap of the semiconductor. As temperature increases, the thermally generated intrinsic carrier concentration increases with a rate that is increasing with decreasing bandgap energy. The importance of the value of n_i is, that for sufficiently high temperatures the intrinsic carrier density exceeds the dopant density (i.e. the extrinsic carrier density), such that the junctions forming the device are washed out and the device fails. In this sense the intrinsic temperature T_i, at which the intrinsic carrier density equals the dopant density N_b of the device substrate, is a measure for the upper limit of the operation temperature of devices isolated by reversed biased junctions. As an example, the intrinsic temperature for Si, GaAs and SiC as a function of the substrate doping concentration is plotted in Fig. 1.

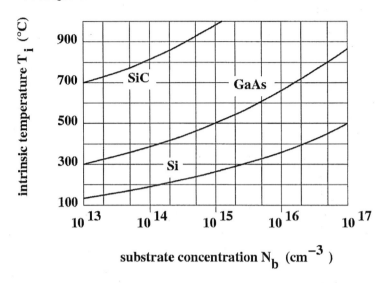

Fig. 1. Intrinsic Temperature T_i for Several Semiconductor Materials

Assuming a substrate doping concentration of $N_b = 10^{15}$ cm^{-3}, due to Fig.1 the intrinsic temperature for Si, GaAs and SiC is about 260°C, 500°C and 980°C respectively. Of course, choosing a higher doping concentration will increase the intrinsic temperature. For example, choosing a substrate doping concentration of 2×10^{16} cm^{-3} will increase the intrinsic temperature for silicon to almost 400°C. However, the dopant density is in turn limited by reverse breakdown considerations at the PN-junctions. Beyond a certain point the main alternative is to use a material with a wider bandgap, as illustrated in Fig. 1.

Keeping in mind Table 1, we can see that typical applications for wide bandgap semiconductors (WBS) in automotive electronics and microsystems are found in the combustion chamber or the exhaust system. Due to the strict low cost requirements in the automotive industry, a silicon-based solution would be of great advantage for all applications with device operating temperatures and subsequently junction temperatures of up to 300°C. In order to achieve this, it is not sufficient to increase the substrate doping concentration, because, besides the intrinsic temperature, there are several other physical parameters, which are strongly affected by the operation temperature of the electronic devices. Some of these parameters will be discussed below, using a CMOS-inverter as an example.

Fig. 2 shows a cross section of a CMOS-inverter consisting of a balanced pair of N- and P-channel transistors, achieved by fabricating the devices in wells of opposite impurity. Electrical insulation between the two devices is achieved by reverse biasing of PN-junctions between the two active devices, an approach referred to as junction insulation.

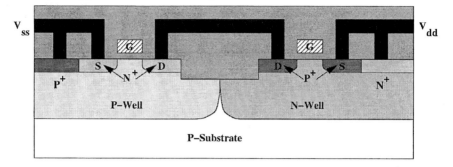

Fig. 2. CMOS-inverter

High-temperature operation of standard bulk CMOS integrated circuits is limited to approximately 200°C because of:

- drift of the **threshold voltage**,
- degradation of the **charge carrier mobility**,
- increase in **off-current**, and
- thermally-induced **latchup**

The influence of the first three effects on the transistor current is shown in the transfer curve of the transistor given in Fig. 3.

Threshold Voltage V_T

The threshold voltage of a MOS transistor is given by the potential difference between gate and source at which the channel between source and drain is formed. In good approximation this voltage is proportional to the width of the space charge region underneath the gate electrode. Due to the increase in n_i with increasing temperature, the width of the space charge region and therefore the threshold voltage will decrease. Simultaneously there will be an increase in the subthreshold swing S with increasing temperature. S is inversely proportional to the slope of the transfer characteristic at $U_{GS} < V_T$ (see Fig. 3). Decrease in V_T as well as increase in S will result in too high an off-current, which is the drain current at $U_{GS} = 0$ V. There are some technological measures to increase the threshold voltage. However, when operating at lower temperatures the term $U_{GS} - V_T$, which is a measure for the on-current of the transistor, might be too low.

Charge Carrier Mobility

Due to the decrease of carrier mobility with increasing temperature the current in the active transistor regime decreases as can be seen in Fig. 3. This current reduction will have detrimental effects on the electrical characteristic of analog and digital circuits.

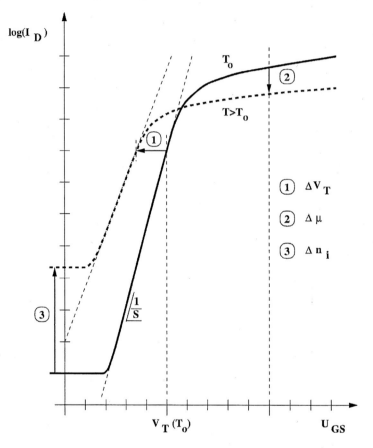

Fig. 3. Transfer Characteristic of a N-channel Transistor (Arbitrary Units)

Off-current

Besides the above-mentioned subthreshold current, the off-current consists of charge carriers generated within the space charge regions of the transistor (generation current) and the surrounded neutral transistor region (diffusion current). While the generation current is proportional to n_i and the total volume of the space charge region, the diffusion current is proportional to n_i^2 and the volume of the adjacent neutral semiconductor material roughly given by the device area and the so called diffusion length of the minority carriers. Therefore, both currents increase exponentially with temperature. Furthermore, shrinking device dimensions will scale both current contributions. However, due to the n_i/n_i^2 relationship the diffusion current will dominate at temperatures above approximately 130°C. Thus, the question arises: how to scale the neutral device volume? The answer will be SOI.

Latchup

As can be seen in Fig. 2 there exist two parasitic bipolar transistors, which are structurally inherent to bulk CMOS: an NPN, given by the source/drain of the N-channel transistor as the emitter, the P-well/P-substrate as the base and the N-well as the collector, and a PNP, with the source/drain of the P-channel transistor as the emitter, the N-well as the base and the P-well as the collector. Due to the fact, that the base of the NPN forms the collector of the PNP and vice versa, there might be a strong interaction between these two transistors forming an electrical switch (latch). Under certain conditions the operation of the bipolars can dominate the behaviour of the total circuit. In particular, switching the bipolar circuit from its normally high impedance state into its low impedance state will result in a low impedance path between the power supply rails. This effect is called latchup [3]. The low impedance state could cause a circuit malfunction. Increasing the temperature will increase the current gain of the bipolar transistors as well as the leakage current within the CMOS structure. Both effects will result in a higher latchup probability. Scaling of the device geometries, as recommended to reduce the off-current, makes things worse due to the resulting increase in current gain of the bipolar transistors. Latchup can also lead to circuit destruction by thermal drifting. An increase in device temperature caused by thermally generated leakage currents leads among other things to an increase in the triggering current that may switch on the parasitic NPNP-structure. Positive feedback will then lead to destruction of the device structure.

3   SOI Structure

The maximum operating temperature of a silicon-based electronic system may be increased by suitable techniques concerning the circuit design, the layout of the circuit and the process and material technology. Within the framework of this article we will restrict ourselves to modifications of material technology in order to improve device characteristics concerning latchup, leakage current and threshold voltage variation.

To avoid latchup we have to prevent the bipolar action within the CMOS structure, i.e. there should be no current flow possible from the N-channel to the P-channel device and vice versa within the silicon substrate. Thus, we the devices have to be laterally (from device to device) as well as vertically (between the devices and the underlying substrate) dielectrically isolated. For this reason we need a prime grade silicon layer on top of an insulating layer as a starting material. Such a structure is called SOI (Silicon On Insulator). Usually the insulator, which is silicon dioxide, is located on top of a silicon substrate, resulting in a sandwich consisting of the device layer, the buried oxide (BOX), and the silicon substrate. Depending on the thickness of the device layer, the lateral isolation may be carried out by trench etching and filling with an insulating material (Fig. 4a) or just by local oxidation (Fig. 4b).

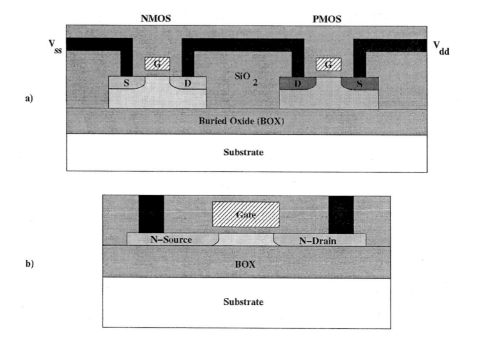

Fig. 4. SOI Structures: a) Thick Film SOI, b) Fully Depleted SOI

As long as the device layer is thinner than the diffusion length of the minority carriers, which is roughly about 1 µm to 100 µm, the contribution of the diffusion current to the leakage current will be reduced. The thinner the device layer, the smaller the generation volume and therefore the leakage current. In the extreme case the device layer may be thinner than the space charge region accompanied with the device, such that the depletion width will not change as a function of temperature. Therefore, changes in threshold voltage would be greatly reduced. To use this advantage device layer thickness of less than 100 nm is necessary. Devices that fulfill this condition are called fully depleted SOI devices. Actually, the choice of the device layer thickness as well as the BOX thickness depends on the application of the electronic system. The diagram shown in Fig. 5 gives some examples of application for SOI structures [4]. The essential methods for the fabrication of SOI materials will be introduced below.

4 SOI Materials

The original driving force behind SOI technology was the need to produce radiation-hardened ICs using an alternative to expensive silicon on sapphire (SOS) material. Thus, several fabrication methods were investigated [5], out of which two succeeded: Separation by Implantation of Oxygen (SIMOX) and the Bond and Etch back technology (BESOI). In order to increase product quality and to reduce wafer-manufacturing cost of the SOI material, based on these two methods, so called layer transfer techniques were developed. These methods are described below.

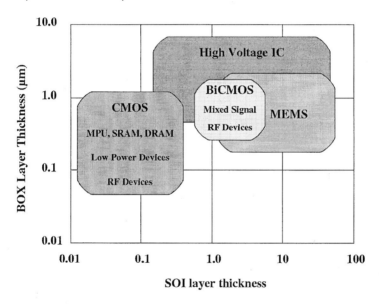

Fig. 5. SOI Layer Configuration for Various Applications

Separation by implantation of oxygen (SIMOX [6]). A high dose of oxygen ions (2×10^{18} cm^{-2}) is implanted into a silicon substrate at energies of about 200 keV. The wafers are then subjected to a high-temperature anneal (1300°C to 1400°C) for several hours to form a buried layer of silicon dioxide (BOX). During annealing the crystalline quality of the silicon layer over the BOX layer will be restored. This standard SIMOX process produces top silicon and buried oxide layers of about 200 nm and 380 nm, respectively.

A modification of the SIMOX process is the PIII process (plasma immersion ion implantation). In PIII, the wafer is immersed in a plasma from which oxygen ions are extracted and accelerated through a high voltage sheath to be implanted into the wafer. Since the entire sample is implanted simultaneously, the implantation time is independent of sample size, which is a prerequisite for a low-cost SOI material. However, since there are no filtering and collimating elements in a PIII reactor, contaminants in the plasma are co-implanted, which may reduce carrier lifetime and therefore degrade electrical device behaviour.

Bond and etch back SOI (BESOI [7]). In this approach, two silicon wafers, both with thermally grown oxide layers, are bonded together using Van der Waals forces. The combined oxide layer formed by the two bonded SiO_2 layers will become the BOX. Subsequently, annealing at temperatures above 800°C in an inert or oxidizing atmosphere is carried out in order to form covalent bonds and thus strengthen the bonding between the two surfaces. After wafer bonding almost the entirety of one wafer is then removed by an assortment of methods – grinding, etching, polishing – such that the remaining silicon on the thinned wafer is now situated above the BOX. The uniformity of the thinned wafer is within 10-30%. Therefore, the minimum device layer thickness will be limited to about 1 µm. Better uniformity can be achieved with a chemical etch stop. Typically, the bonded wafers have thicker silicon and buried oxide layers of higher quality compared to SIMOX material. However, the BESOI process requires two starting wafers, adding to the expense of the process as compared to the one wafer for SIMOX fabrication.

Neither SIMOX nor BESOI meets all requirements necessary for SOI materials suitable for mass production of a multitude of electronic systems. The criteria for this are:

- The quality of the device active layer should be equivalent to that of the mono-crystalline silicon substrate.
- The quality of the BOX should correspond to that of thermally grown oxide.
- To reduce costs, the process should need effectively just one wafer for the fabrication of a SOI wafer.
- In addition, the process should be suitable to adjust the thickness of the device active layer as well as the BOX layer.

Especially, the fabrication of device active layers thinner than 100 nm with high uniformity across large wafer diameter should be possible.

The layer transfer technique, which will be discussed now, has the potential to significantly reduce wafer production costs while ensuring high-quality SOI wafers. A conceptual view of this technique is shown in Fig. 6 [8]. The process starts with bonding of the so called handle wafer B (which serves as a substrate for the SOI wafer) to the so called seed wafer A, which is treated to form an oxide layer, the device active layer and the so called splitting layer. Then wafer A is split-off using a certain procedure (wafer separation) to be discussed below. After that, the handle wafer with the SOI layer on top gets the final treatment to complete the SOI wafer fabrication process. The split-off seed wafer is processed for reuse (either as a handle or seed wafer), which results in significant cost reduction of the total process. Based on this method, there are various fabrication processes for SOI wafers, which have some common elements: the methods for formation and treatment of the device active layer and the split-off layer on wafer A, and the method for splitting off wafer A after bonding and formation of the SOI layer.

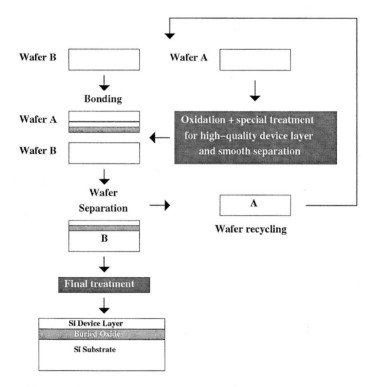

Fig. 6. Conceptual View of the Layer Transfer Technique

Smart Cut Technology [9]. This technique combines ion implantation and wafer bonding. First, seed wafer A is oxidized to form what will become the BOX layer of the SOI structure. Then the splitting layer is formed by ion implantation of hydrogen at relatively high doses ($5 \times 10^{16} \text{cm}^{-2}$) into wafer A.

This implantation induces the formation of an in-depth weakened layer (accumulation of hydrogen cavities) located at the mean ion penetration depth. The process continues with bonding seed wafer A to handle wafer B using Van der Waals forces. After that, a two-phase heat treatment of the bonded wafers follows. During the first phase at about 500°C, the hydrogen cavities merges causing seed wafer A to split into two parts: a thin layer of single-crystalline silicon remaining bonded to handle wafer B and the rest of seed wafer A, which can be recycled for re-use as handle wafer. The second high temperature treatment (> 1000°C) of wafer B, aims to strengthen the chemical bond. After splitting, the layer on top of wafer B exhibits a microroughness, which makes chemo-mechanical polishing of the surface necessary.

Wafer bonding in conjunction with layer transfer technique enable different materials to be associated to form multilayer substrates for microsystem applications. This technique is especially interesting for expensive WBS materials such as GaAs, GaN, SiC, diamond and a number of insulating and conductive bonding layers. These technologies, which may open the way to real high temperature applications, are under development with different grade of maturity.

Epitaxial Layer Transfer technique (ELTRAN [10]). In contrast to the other procedures already discussed, in this method the device active layer is deposited epitaxially. First, porous silicon is formed at the surface of the seed wafer by means of anodisation. The anodisation involves passing a current through aqueous hydrofluoric acid with the seed wafer used as anode in order to form microscopic pores on the surface of this wafer. The structure of the porous silicon can be easily controlled by the current density. Increasing the current in the middle of the anodising process creates a two-layered porous silicon: a first porous layer (on top) with low porosity in order to increase the quality of the epitaxial silicon, which will be deposited later on, and a second layer, the porosity of which is optimised with respect to the splitting process. Hydrogen pre-baking in conjunction with pre-injection of silicon atoms out of a gas phase causes the pores in the porous silicon surface to close up, resulting in a smooth surface. A single-crystal silicon layer is now epitaxially grown on top of the smoothed layer. Next, the surface of the epitaxially silicon layer is thermally oxidized. The resulting oxide film will become the BOX layer of the SOI wafer. The seed wafer and handle wafer are now bonded and thermally treated, reinforcing bonding strength. A water-jet technique is used to split-off the seed wafer in the middle of the porous silicon layer of the bonded wafer. Splitting takes place parallel to the wafer surface close to the interface of the two-layered porous silicon layer. Roughly speaking, the first porous silicon layer ends up on the handle wafer (i.e. the SOI wafer) side, while the second layer remains on the seed wafer side. While the split-off seed wafer is processed for reuse, the final treatment of the SOI wafer consists of selective etching of the residuary porous layer and hydrogen anneal of the sample to improve surface quality.

5 Conclusion

The automotive industry drives the development of electronic systems operating at temperatures higher than the mil-specified limit of 125°C. Due to the strict low cost requirements in the automotive industry, only silicon-based solutions are seen to be promising for mass production. Choosing an appropriate circuit design and process technology, bulk CMOS may operate reliable up to 200°C. This limit is pushed to temperatures around 300°C to 350°C using SOI techniques. For the use at even higher temperatures, a new semiconductor technology must be introduced. The physical and chemical durability of SiC combined with the rapid progress in device and material technology suggest that SiC is likely to play a key role in future automotive electronic systems.

References

[1] J. Marek: "Microsystems for Automotive Industry", in IEDM Proc. 2000, pp. 3-8.

[2] W. Wondrak: "Physical Limits and Lifetime Limitations of Semiconductor Devices at High Temperatures", Microelectronics Reliability 39 (1999), pp. 1113-1120.

[3] R. Troutman: "Latchup in CMOS Technology – The Problem and its Cure", Kluwer Academic Publishers, 1986.

[4] http://www.canon.com/eltran/soi

[5] P. Seegebrecht: "SOI Technologien", GME Fachbericht 4 "Mikroelektronik", VDE Verlag, 1989, pp. 97-102.

[6] K. Izumi et al.: "SIMOX Technology for CMOS LSI's", VLSI Symp. Tech. Digest, 1982, p. 2.

[7] J. Haisma et al.: "Silicon-on-insulator wafer bonding – wafer thinning technological evaluations", Jpn. J. Appl. Phys. vol. 28, no. 8, 1989, pp. 1426-1443.

[8] O. Takashi: "At long last, SOI wafer market on the move", Solid State Technology, Feb. 2001, pp. 62-65.

[9] M. Bruel: "Silicon on Insulator Material Technology", Electronic Letters, Vol. 31, 1995, pp. 1201-1202.

[10] K. Sakaguchi, T. Yonehara: "SOI wafers based on epitaxial technology", Solid State Technology, June 2000, pp. 88-92

2.5 Microsystems Packaging for Automotive Applications

E. Jung[1], M. Wiemer[2], V. Grosser[1], R. Aschenbrenner[1], H. Reichl[3]

[1]FhG-IZM Berlin
 E-mail: erju@izm.fhg.de

[2]FhG-IZM Chemnitz

[3]TU Berlin

1 Introduction: Role of MEMS in the Automotive

Micro Electro Mechanical Systems (MEMS) have gained a crucial role in the automotive industry over the last decade. Initially for sensing various environmental, motor and movement situations, today actuators become more part of the automotive as well. For sensing, early fields of application included pressure sensors and acceleration sensors. Today, a large number of safety and comfort features rely on sensors [1]:

Pressure sensors	tyre pressure, injection pressure, exhaust gas pressure, manifold inlet, combustion pressure...
Acceleration sensors	airbag ignition, smart suspension, ...
Gyroscopes	roll over detection, ESP,...
Optical sensors	rain drop detection, seat occupancy, auto-dimming mirror, ...
Flow meters	air mass flow
Shear sensors	oil quality
Temperature sensors	air inlet temperature, exhaust gas temperature
Electric and magnetic field sensors	oil quality, child seat detection, ABS, oil sump level

Table 1. Fields of Application for Sensors Using MEMS Technology (Table not Exhaustive)

For actuators, new fields for comfort, ecology and safety arise like

- Head lamp actuators for Xenon lamp regulation
- Direct injection using piezoelectric valves

Not at last, integral solutions with sensor/acutator systems will have a significant role in the future of automotive MEMS, e.g.

- Auto cruise control (radar distance sensor coupled to a speed control/brake)
- Uncooled IR Sensors for night drive collision warning coupled to brake

While hybrid electronics paved the way for the electronic integration into the automotive, MEMS/hybrids will play a more and more dominant role in the future. Packaging innovations here are crucial to fulfil the stringent requirements for the automotive market like:

- High reliability of function
- High availability of function
- Ease of manufacturing
- Low cost devices

Aside, miniaturization is a part of the development goals not to be neglected, as it is expected that future electronics will comprise a significant part of the automotive value. To integrate a high degree of electronic functionality into the already densely packaged electronic compartment of the car requires miniaturization at its best. The last decade has seen tremendous increase in functionality while decreasing size and volume (Fig. 1).

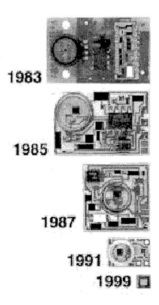

Fig. 1. Miniaturization Evolution of a Pressure Sensor [2]

According to NEXUS, the value of the microsystems content in the automobile will increase from ~1.94 Mrd. Euro in 2001 to ~2.52 Mrd. Euro in 2005 at an astounding 16.9% annual growth rate.

2 Role of MEMS Packaging

MEMS packaging fulfils a crucial role within the function of the MEMS device. Besides the electrical interconnection, the package provides mechanical protection, media separation or coupling (e.g. pressure), signal conditioning, etc., it must enable the resulting package to be manufactured and assembled at low cost. Also, high reliability requirements even under harsh conditions (e.g. oil sink with extreme temperature differences and aggressive ambient) must be fulfilled in order to be accepted in the automotive market.

Today, packaging relies either on using well established package types like TO-headers, butterfly or ceramic packages, or – applicable only for very high volume products – customized package developments. Wafer level packages, as being developed in the microelectronic industry, are currently finding their niche and potential wide spread use also for the packaging of MEMS devices. This can incur substantial cost savings in the future, as seen in the microelectronics industry as well.

3 Sensor Packaging Options

3.1 Wafer Level Preparation vs. Wafer Level Packaging

Basis for the large majority of moving MEMS devices like accelerometers and gyros is the process of structure protection while still on wafer level. Therefore, a structured "cap" wafer is used (typically consisting of pyrex glass or silicon) to cover the delicate structures. This capping wafer is aligned to the device wafer and bonded to this via one of the following technologies (order according to share of use):

- Anodic bonding
- Glass frit bonding
- Adhesive bonding
- Silicon direct bonding
- Soldering

Each of the technologies has their distinctive advantages and disadvantages, so the choice depends on the product and product's requirements in operation.

Besides adding just the protective layer, alternative technologies like Shell case provide in addition to hermeticity and mechanical separation the terminal contacts ready to be mounted in SMD fashion. Here, the technological evolution has significantly increased the reliability of these devices as well [3].

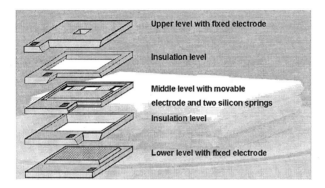

Fig. 2. Accelerometer Manufactured by Sequential Layering of Wafers

While in these cases, the mentioned technologies provide only mechanical and environmental protection (e.g. hermetic enclosure of the moving device), other approaches use this necessary process step to add functionality to the structure. Here, the layers providing the function of the device are added sequentially via anodic or adhesive bonding. E.g. accelerometers or pumps have been realized in this fashion [4].

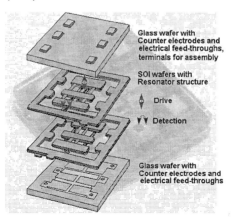

Fig. 3. Functional Layering with SMD Ready Contacts

A next step of evolution here is to simultaneously provide protection, functionality and wafer level applied SMD contact structures to the device [5]. Wafer level packaging including the terminals is expected to be one of the major technology trends in the near future of MEMS packaging.

3.2 Single Chip Preparation

The vast majority of devices is packaged using well established technologies known from the microelectronics industry for several decades. With movable MEMS devices, the first step is capping while still on wafer level. This process is part of the manufacturing process in the MEMS fab. With less delicate or media sensitive devices like field detectors or thick film sensors, such rigorous protection is not necessary. After fabrication, the wafers are diced using standard dicing equipment. Then, they are singulated and packaged in individual housings, e.g.

- Metal
- Ceramic
- Plastic

packages.

Metal housings are e.g. TO cans, butterfly packages (as used in the optoelectronics industry) or may also be machined directly for the application. While the TO approach can rely on a well established infrastructure, the situation is less easy available with butterfly packages and only available for specific products with the machined housings. TO cans lend themselves to volume production even for consumer goods and provide excellent mechanical and environmental protection (hermeticity). However, the footprint of the device is non-optimal both with respect to miniaturization and SMD manufacturing.

Fig. 4. Multi-Die TO Can Package for Acceleration Sensor with Read-out Circuitry [6]

With in-plane assembly onto a carrier substrate, multi-die TO cans can be realized (Fig. 4). When volumetric packaging is an option, flex substrates can be used to realize 3D packages(Fig. 5).

Fig. 5. Volumetric Packaging of a MEMS Subsystem Into a TO Can Using Flex

Sealing of these devices uses either adhesives (non-hermetic) or – more usually - weld sealing which provides a hermetic enclosure and is e.g. especially useful for pressure sensors with a well defined pressure in the cavity [6].

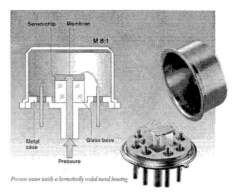

Fig. 6. TO Packaged Pressure Sensor (Courtesy BOSCH)

For butterfly packages and machined housings, dedicated solutions for welding, soldering and e.g. glass lid attach exist.

Ceramic housings proved a much larger cavity than TO cans and are as well available form the microelectronics infrastructure in volume numbers. Their special advantage comes into play, when optical sensors (pyrometers, cameras) are to be packaged using glass lids. The device itself is mounted in standard wirebond technology and the cavity is closed with a glass lid using glass frit bonding or adhesive bonding (non-hermetic). A special feature of the ceramic is to serve as a routing substrate as well, providing solderable and wirebondable surfaces with mid-density routing capabilities.

Fig. 7. Pressure Sensor with Metal Lid (Backside) and Pressure Inlets, Ceramic Substrate [7]

For modular applications using MEMS, a recently commercialised technology "MATCH-X" lends itself to package MEMS devices into ceramic housing while allowing electrical buses, pressure, optical and fluidic access into the package [8] (see Fig. 8).

Sealing techniques used with ceramic substrates are soldering, glass frit bonding and adhesive bonding (non-hermetic).

Fig. 8. MATCH-X Modular MEMS Packaging Using LTCC Ceramic

Plastic packages are favoured by the microelectronics industry due to their manufacturing efficiency and low costs. For the packaging of MEMS devices, they don not lend them easily. Stress induced due to the electronic molding compound (EMC) process may influence sensor characteristics, media access (e.g. pressure of sampling gas) is difficult to achieve, the number of devices to be packaged is several orders of magnitude smaller than those of microelectronic devices of the same size. Therefore, pre-molded substrates have found widespread use in the manufacturing of MEMS packages. Here, the devices are mounted similarly to ceramic packages. Lid attach using injection molded caps provide mechanical protection while allow media access to the device. Instead of lid attach, for several applications like pressure sensors a low modulus gel is used to protect the wirebonds and hydrostatically transfer the external pressure to the device (Fig. 9).

Fig. 9. Pre-Molded Package Filled with Low Modulus Gel

In the last years, technologies have begun to emerge that allow the low cost EMC molding process to be applied even to MEMS packaging. Here, media access is provided via compliant nubbins within the mold form or encapsulating with the device sacrificial spacers.

With this approach, MEMS find their way into "system-in-package" applications, which provide a subsystem functionality e.g. in the form of a surface mountable BGA or flex tape carrier (Fig. 10).

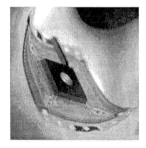
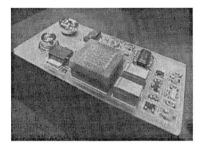

Fig. 10. Flex Substrate for MEMS Packaging with Overmold and Media Access Cavity

Fig. 11. MATCH-X Modular MEMS Package Using Organic Laminate

Especially for low cost applications requiring a modular approach (e.g. different sensors, different bus systems, ...), the three dimensional "Match-X" concept is moving from the ceramic housing to organic laminate substrates. Significant savings and competitive cost structure to discrete system assembly are expected.

Application specific packages (i.e. packages designed for one specific device and purpose [9]), as they are known in e.g. the medical or optoelectronics industry are not simply applicable in the automotive sector, since the volumes involved and the price obtained with those ASIPs is typically not compatible with automotive requirements.

Outlook – Functional Packages

Conventionally, packaging of the devices is a separate process step providing a structure which can be handled during the subsequent manufacturing sequence and provides the functionality required. However, for high volume products and for optimised performance, an entire (sub-) system may be designed from the beginning to provide highest cost efficiency, highest function density and optimum performance. Such devices are "functional packages" which not only encase the devices's function but add their own functionality to maximize the overall performance. So far, functional packages find their applications in the consumer industry (e.g. ink jet cartridges) and medical industry (blood sensors), e.g. in throw-away products. However, with "platform concepts" "re-use of technology" over the various automotive models, the automotive industry can take advantage of the concepts developed for these ultra-low cost products.

Annex – Roadmap of MEMS Packages

The following figures show the expected technological evolution for MEMS packages as derived from the perspective of

- increased or maintained reliability
- higher packaging density
- higher modularity
- lower cost

Single Device Packages

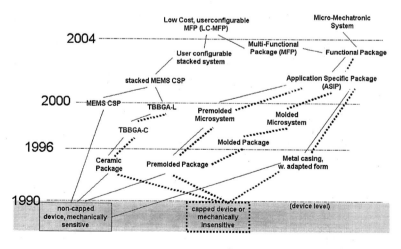

Fig. 12. MEMS Packaging Roadmap for Single Chip Packages

Wafer Level Packages

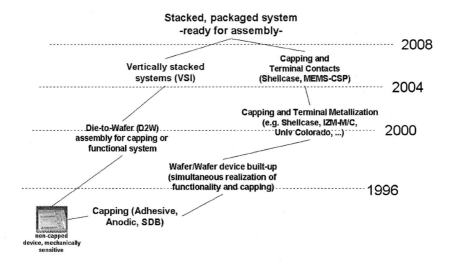

Fig. 13. MEMS Packaging Roadmap for Wafer Level Packages

References

[1] Krueger et al., MST News 01/01.

[2] Schuster et al, MST News 02/99.

[3] http://www.shellcase.com/pages/products.asp

[4] T. Gessner et al.: "High precision acceleration sensor in silicon", Proceedings to Sensor 95, Nuremberg, S. 409.

[5] M. Wiemer et al "Application of high and low wafer bonding processes for bulk micromachined components", Proc. of the 5th Int. Symp. on Semicond. Wafer Bonding: Science, Technology and Applications V; 1999, Honolulu, Hawaii.

[6] http://www.europractice.bosch.com/en/foundry/index.htm

[7] http://www.novasensor.com/catalog/NPC_series.html

[8] H. Reichl, V. Grosser, "Overview and development trends in the field of MEMS packaging", IEEE MEMS 2001, Interlaken, 2001, pp. 1.

[9] http://www.infineon.com/products/chipcds/portfol/biometr/introduction.htm

2.6 Micro-Mechatronics in Automotive Applications

F. Ansorge, J. Wolter, H. Hanisch, V. Grosser, B. Michel, H. Reichl

Fraunhofer Institute Reliability and Microintegration
Micro-Mechatronic Center
Argelsrieder Feld 6, 82234 Wessling, Germany
Phone: +49/8153/9097-500, Fax:+49/8153/9097-511
E-mail: ansorge@mmz.izm.fhg.de

Micro-Mechatronics & System Integration is a multidisciplinary field, which offers low cost system solutions based on the principle of homogenizing system components and consequent elimination of at least one material, component or packaging level from the system. These system approaches show, compared to the existing solutions, a higher functionality, more intelligence and better reliability performance. The number of interconnects necessary to link a motor, sensor, or an actuator to a digital bus decreases, as the smartness of the devices increases.

The paper presents system solutions and manufacturing technology for mechatronic systems. To reduce package volume, advanced packaging technologies such as Flip Chip and CSP are used, for increased reliability and additional mechanical functionality encapsulation processes as transfer-molding, a combination of transfer- and injection molding or a modular packaging toolkit based on LTCC is selected.

The first system is a multi-chip-module for motor control in power windows or electric sun-roofs. For the module at least three interconnection layers were eliminated using novel concepts like intelligent leadframes and integrated plugs. The package resists harsh environment as present in automotive applications.

Another mechatronic package called TB-BGA or StackPac respectively, is a 3-D solution, containing a pressure sensor and the complete electronics necessary for control and data transfers. This package involves CSP-technologies on flex, with an increase of functionality per volume unit by direct CSP-to CSP mounting, eliminating substrate layers and large signal distances of unamplified signals from sensor output.

Mechatronic packages will be discussed in detail in this paper. Especially, cost, reliability performance and according "Design for Reliability" show the potential of micro-mechatronic solutions in automotive and industrial applications.

1 Introduction

The development of electromechanic assemblies has so far been carried out mainly by different manufacturers all following their own individual methodology. Assembly and fine adjustment of the individual components and their integration into a complete system normally only takes place at the end of the manufacturing process.

Due increased system intelligence, many electronic systems comprise an additional control unit which receives information from different sensors and selects the individual electromechanic actors. These central control units need a lot of space and are fairly inert due to the long transmission paths of the electric signals. On top of that, the wiring of all the different components involved in the assembly process can be a complex and expensive business; it is definitely a major headache for mobile applications.

Sub-systems on the other hand, which mechatronically integrate both actor and sensor facilities as well as all information processing (software) in the component itself, deliver an optimized use of existing space and higher speeds of data transmission between actor/sensor functions and µ-processor technology. All this results in substantially higher performance capabilities. This would reduce transmission distances and data volumes. It would also direct the individual systems towards greater independence and autonomy. Both arguments are crucial for optical devices

The overall system is characterized by the communication of the individual sub-systems both with each other and the central control unit (main controller)(see Fig. 1). This process can be conducted via collision-insensitive bus systems. A field bus system connecting all sub-modules with the main controller or a (thus defined) master module in the form of a ring or a string is already enough to ensure the free flow of all information.

The basic sensor data were already processed and converted in the sub-module. Only the data required by the control logic system is transmitted to the main control unit. As a consequence, you have electromechanic systems with "subsidiary intelligence" – a mechatronic system.

This multi-directional feedback mechanism results in a dominant technology, based on subtle analysis of the functions needed and optimum separation of functionality of the parts involved. Mechatronics is defined here in this sense as the application of intelligent sub-modules. A more comprehensive definition would define it along such lines as the following: Mechatronics is the synergistic integration of mechanical engineering, electronics and complex control logic into development and working

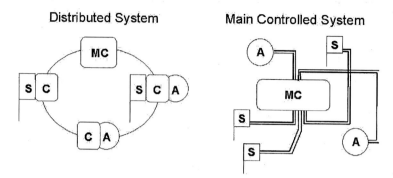

Fig. 1. Juxtaposition of a Sub-system with Bus Technology (Left) and a Conventional Control Unit (Right). (MC Main Controller; C Controller; A Actor-Mirror; S Sensor)

procedures, resulting in mechanic and optic functionality and integrated algorithms [1], [2].

2 Basic Conditions for the Development of Mechatronics

The development of mechatronics is a vital key for the future of electronics. Ever more complex applications require the processing of ever increasing data volumes: a corresponding increase in flexibility and functionality is the only possible response developing micro-mechatronics. Synergistic co-operation between the individual departments concerned is also an indispensable condition. Electronic and mechanical simulation and designs must be compatible for a proper realization of the thermomechanical concept (see Fig. 2). The package developed fits to an optimum place within a macroscopic system of mechanical parts, which are especially designed to work with electronics as intelligent sub-system. It is of equal importance to generate a software map of all requirements and functions, if at all possible already at the planning and development stage; this is the ideal way to shorten development times while optimizing the system eventually produced. To yield this added functionality within one package the effort of specialists from various fields of technology as physics, chemistry, electrical engineering, microelectronics, software design, biology, medicine etc. are needed.

2.6 Micro-Mechatronics in Automotive Applications

Mechatronic Design

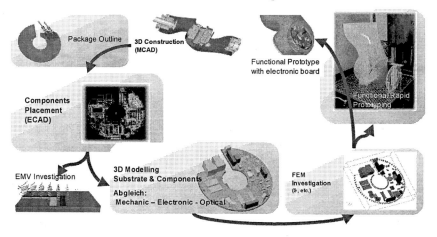

Fig. 2. Mechatronic Design Procedure

2.1 Packaging Concepts and their Aspects on Polymer Materials for the Use in Automotive Applications

Focusing on micro-mechatronic applications there are a few additional demands to encapsulate commonly used for packaging. The encapsulates need a wide range of temperatures, a high resistance against harsh environment and the integration of moving and sensing elements without losing package functionality. Additionally for the use of micro-mirror devices clear compounds are crucial [7].

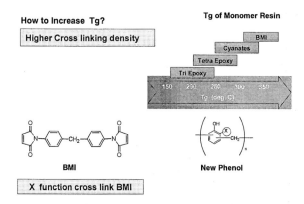

Fig. 2. High Tg Materials [10]

Typical materials used for microelectronics encapsulation are epoxy resins, where the chemical basis is a multifunctional epoxy oligomer based on novolac resins. These materials do have Tg's beyond 200°C and so they have the potential for short term / high temperature application. Fig. 3 will give an impression of developments for future high Tg materials by Shin-Etsu Chemical Co., Ltd [10].

The evaluation of encapsulates for optimized long-term stability is one of the topics the micro-mechatronics center is focusing on. Clear compounds as potting or molding material are also already available, but due to the lack of fillerparticle there the CTE is much higher than for standard materials. This needs more sophisticated selection of package outline and arrangement of the components.

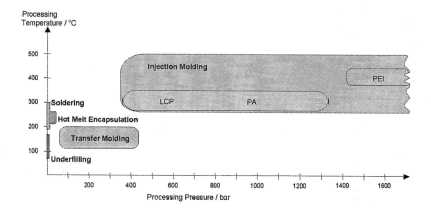

Fig. 3. Process Window of Thermoplastic and Thermoset Materials

Further potential for micro-mechatronics are in the use of advanced thermoplastic materials. In the past they have been used rather for electronics housing than for the direct encapsulation of microelectronics.

The use of thermoplastic materials not only allows the integration of additional mechanical functionality (plug housing, clip on mounting, ...) simultaneously with microelectronics encapsulation, but these materials do not cross-link and thus do have a high potential for recycling.

Research of Fraunhofer IZM-MMZ is performed in the fields of direct encapsulation of microelectronics using thermoplastic polymers, thermoplastic circuit boards a.k.a MID-devices for both, Flip Chip and SMD components. Combination of different types of materials, tailored for the application, is a key for highly reliable modules.

2.6 Micro-Mechatronics in Automotive Applications   81

These applications are also capable for connecting the package geometry with mechanical functions. Only thermoplastic materials are suited for the creation of scoop-proof connectors or clip-on-connections. Their suitability for the process of direct encapsulation, is subject to a number of conditions. In order to preserve a multifunctional package, both technologies would have to be joined in the forming of two- or multi-component-packages.

Additional areas of research important for the manufacturing of miniaturized mechatronic systems are wafer level encapsulation and the handling of thin and flexible base materials, e.g. thin silicon ICs.

In this case printable epoxy materials seems to be important candidates for improvement of several automotive applications. Table 1 give properties of various commercial printable Semicoat Materials [11].

Properties

Grade		SEMICOAT 400	SEMICOAT400E	X-43-5028FR-1	X-43-5028FR-2	SEMICOAT400FR
Feature		Standard	Low cost type	Modified of SEMICOAT 400E		
		Very Low α type	Regular type	4mm : V-0		1mm, 2mm, 4mm : V-0
		Fine distribution	Normal distribution			
Br / Sb		0 / 0		1 / 1	1.3 / 1	0.9 / 1
Filler content	wt%	79		80		78
Filler Average size	um	4	17		17	
Filler Max. size	um	25	75		75	
ITEM	Unit					
Appearance		Black	Black	Black	Black	Black
Viscosity : BH;No.7; 2rpm	Pa?s	470	220	100	146	820
Viscosity : BH;No.7;20rpm	Pa?s	NA	83	80	102	NA
Thixotropic Ratio		NA	2.7	1.3	1.4	NA
Gelation Time : 150C	s	55	55	58	58	70
Cured Property						
Specific Gravity		1.87	1.87	1.93	1.93	1.87
Glass Transition Temp.	?	155	155	157	155	147
Coefficient of Thermal Expansion : 50-80C	ppm/?	14	13	13	14	14
Coefficient of Thermal Expansion : 200-230C	ppm/?	57	55	51	52	66
Flexural Strength	N/mm2	142	123	145	140	96
Flexural Modulus	N/mm2	16000	12400	14300	13800	9500
Cure condition		90????	150? 2?	90????	150? 2?	90???? 150? 4?
Printing condition (Device temp.)		Room Temp.		Room Temp.		
Storage condition		Below -40?		Below -40?		

Table 1. Properties of Printable Epoxy Materials [11]

The use of thermoplastic hotmelt materials for micro-mechatronic packaging allows a decrease of processing temperature, the use of cost effective low temperature materials [8].

2.2   Modular Construction Systems for Mechatronics

In the course of a project sponsored by the BMBF, the Fraunhofer Institute for Reliability and Microintegration (IZM) developed a modular construction system for sub-systems. Spatially stackable single components include sensor elements, signal processing, actors and bus technology. The components are mounted taking into account the application they are supposed to serve. They produced standard components and can be mounted by soldering, leading to a minimized number of different interconnection technologies.

The standardized dimensions in different size categories enable the user to combine the system components, which are part of the modular construction program, in a most flexible way by defining his own functionality parameters [4]. Connection to the actorics can be effected by a direct integration of the components.

Based on molding-on-flex-technology the StackPac offer its advantages. This technological concept allows the insertion of individual ICs by contacting through either wire bonding or flip chip technology and integration of SMD-components. Multichip modules can be created using different contacting technologies.

Molding the package widens the range of possible package designs from simple patterns to additional functional geometries which are compatible with the modular construction system as depicted in Fig. 5.

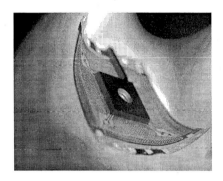

Fig. 4. CavityPack: Molding on Flex Approach to Sensor Packaging

The same technology can be used to design chip size packages (CSP's). The necessary combination with the sensorics can be created by using a CavityPack (see Fig. 4).

The systems have already been used successfully in engineering applications. StackPac's achieve excellent reliability scores, in particular for high-temperature applications, using the mentioned novel encapsulates. This allows a reduction of development and qualification costs by using standard modules, which makes the use of advanced technology in confined spaces economically viable.

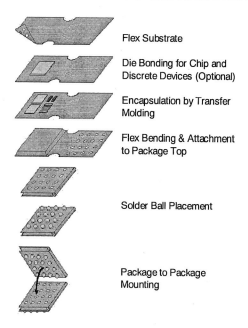

Fig. 5. StackPac Production Process

3 Mechatronic Solutions for the Automobile

3.1 Mechatronic Combination of Power&Logic Electronics and Actuators

In the course of a project jointly run by BMW AG and the Fraunhofer-IZM, an H-bridge circuit with control logic, bus connection and sensor facility has been developed. The use of power ICs required a concept for the heat dissipation of the package. The reduction in the number of interconnecting levels was another purpose of this mission. The power semiconductors have more advantages: in comparison with relays, they use substantially less space and are extremely well suited for the software selection process. This can turn out to be a particularly decisive advantage to conventional solutions, once the 42-V-on-board-network has been installed (See Fig. 5).

Thermal management functions are fulfilled perfectly by the heat sink comprising the inserted copper lead frame and the engine which has been mounted right at the bottom side of the package. The motor block has the effect of an active cooling element. This integrated thermal concept allowed the substitution of the conventional trigger system by spatially optimized electronics (See Fig. 6). The high degree of integration resulted in a reduction of spatial requirements by about 50% (See Fig. 7).

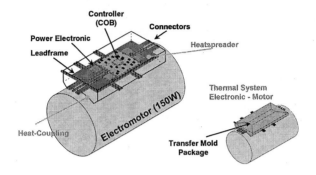

Fig. 6. H-Bridge Controller

Windowlift Mechatronics - Today and Tomorrow

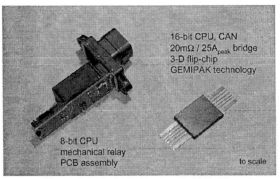

GEMIPAK - A joint R&D project of BMW, Brose, Fraunhofer IZM, and Motorola
Funded by BMBF / VDIVDE-IT

Fig. 7. Gemipak Controller

3.2 Micro-Mechatronic Applications: Micro-Mirrors Used for Front Light Systems

A new application for micro mirror devices will be the front light system for automobiles. A brand-new functionality is the control of the lamp reflector while driving along curves and subject to the view-direction of the driver (or by speech control). For this a part of the reflector mirror from the front headlights can be controlled so that the roadside which is aimed at by the driver will be additionally illuminated short-time. A detailed report is give in [11] in this issue.

4 Outlook

According to intensive research studies, a large part of what is today still a vision of the future we also gave you a glimpse of that in our paper will be converted into reality by micro- machine technology already in the period between 2002 and 2007: the integration of sensorics, actoric functions and controllers will make it possible.

	Powertrain management		Safety		Comfort, convenience, security	Diagnostics and monitoring	Information / communi - cation
	Engine	Trans-mission	Passive	Active			
Shares of Sensorics			++	++	++	++	+
Miniaturization	++	++	++	++	++	++	++
Modularity	+	+	+	+	+	+	+
Cost effectives	+++	+++	+++	+++	+++	++	++
Market potential	↗	↗	↗	↗	↗	↗	↗

Table 2. Requirements and Trends for Automotive Applications
(+++ crucial, ++ very important, + important, ☐ optional, ↗ increasing)

Harsh engine environmental and overall vehicle systems complexity coupled with requirements for low costs, size, and weight plus high reliability make the automotive sector. The annual growth in the automotive sector is estimated at around 20 % and will medium-term continue.

Acknowledgement

Parts of this work are funded by the "Bayerisches Kompetenznetzwerk für Mechatronik" part of the "High Tech Offensive Zukunft Bayern".

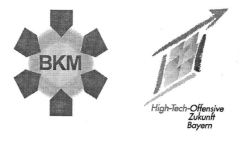

References

[1] Control and Configuration Aspects of Mechatronics, CCAM Proceedings; Verlag ISLE, Ilmenau, 1997.

[2] F. Ansorge, K.-F. Becker; Micro-Mechatronics – Applications, Presentation at Microelectronics Workshop at Institute of Industrial Science, University of Tokyo, Oct. 1998, Tokyo, Japan.

[3] F. Ansorge, K.-F. Becker, G. Azdasht, R. Ehrlich, R. Aschenbrenner, H. Reichl: Recent Progress in Encapsulation Technology and Reliability Investigation of Mechatronic, CSP and BGA Packages using Flip Chip Interconnection Technology, Proc. APCON 97, Sunnyvale, Ca., USA.

[4] Fraunhofer Magazin 4.1998, Fraunhofer Gesellschaft, München.

[5] F. Ansorge; Packaging Roadmap – Advanced Packaging, Future Demands, BGA's; Advanced Packaging Tutorial, SMT-ASIC 98, Nuremberg, Germany.

[6] Toepper, M. Schaldach, S. Fehlberg, C. Karduck, C. Meinherz, K. Heinricht, V. Bader, L. Hoster, P. Coskina, A. Kloeser, O. Ehrmann, H. Reichl: Chip Size Package – Michael A. Mignardi; "From ICs to DMDs", TI Technical Journal, July-September 1998, pp. 56-63.

[7] Long-Sun Huang; "MEMS Packaging for Micro Mirror Switches", Abstract, University of California.

[8] Jeff Faris, Thomas Kocian; "DMD Package – Evolution and strategy"; TI Technical Journal, July-September 1998, pp-87-94.

[9] Fraunhofer Institut Mikroelektronische Schaltungen und Systeme; "Lichtmodulatoren mit mikromechanischen Spiegelarrays" 2001.

[10] Dick Misawa; "Materials for Advanced Packages"; Polytronic Conference 2001, Potsdam Germany.

[11] F. Ansorge, J. Wolter, H. Hanisch, H. Reichl; "New Principle for Near Field Observation"; AMAA Conference 2002, Berlin, Germany.

3 Functions and Applications

3.1 Towards Ambient Intelligence – Obstacle Detection

3.1.1 Overview: Obstacle Detection

A. S. Teuner

Delphi Delco Electronics Europe
Lise-Meitner-Strasse 14, 42119 Wuppertal, Germany
Phone: +49/202/291-3410, Fax: +49/202/291-4140
E-mail: andreas.teuner@delphiauto.com, URL: www.delphiauto.com

Keywords: anticipatory crash-sensing, driver assistance, electro-optical systems, LIDAR, occupant position recognition, RADAR

Abstract

Obstacle detection is a prerequisite for a variety of innovative safety and driver assistance systems such as adaptive cruise control, collision avoidance, anticipatory crash sensing, occupant position recognition or night vision. Depending on the specific application, obstacle detection systems enable the measurement of the position, range, range rate or temperature of objects in the sensor's field of view. Typically, this information will be processed by a subsystem in order to control the vehicle dynamics or to deliver an appropriate warning to the driver that enhances his situational awareness. This overview introduces the motivation for collision warning/avoidance systems and the most common obstacle detection techniques used for exterior and interior sensing. It illustrates typical applications and discusses benefits and drawbacks of each technology. An integration of obstacle detection systems and advanced applications is presented in the form of a concept car built by Delphi to demonstrate the capabilities of future integrated safety systems.

1 Introduction

Tremendous progress has been made since the 1960's with regard to vehicle safety. Early safety approaches emphasized precaution (e.g.: surviving a crash) and focused on passive devices such as seat belts, air bags, crash zones, and lighting. These improvements have dramatically reduced the rate of crash-related injury severity and fatalities. For example, in the United States the fatality rate per hundred million vehicle miles travelled has fallen from 5.5 to 1.7 in the period from the mid-1960s to 1994. However, in spite of these impressive improvements, motor vehicle crashes still account for a staggering 40,000 deaths, more than 3-million injuries, and over $150 billion in economic losses [1] each year. In Europe the annual economic losses are estimated at €100 billion. The amount of damage related to personal injuries or deaths is 56% while the remaining 44% is related to material damage [2]. These numbers emphasize the increased need for obstacle detection systems that operate under all day and night time conditions. Crash statistics and numerical analysis strongly suggest that collision warning systems will be effective. As shown in Fig. 1, crash-related data collected by the U.S. National Highway Traffic Safety Administration (NHTSA) show approximately 88% of rear-end collisions are caused by driver inattention and following too closely. These types of crash events could derive a positive beneficial influence from collision warning systems. In fact, NHTSA countermeasure effectiveness modelling has determined that these types of headway detection systems can theoretically prevent 37% to 74% of all police reported rear-end crashes [2], [4].

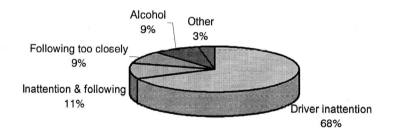

Fig. 1. Causes of Rear - Front Crashes (Source: NHTSA)

The vision of accident-free driving is not new: since the late 1950's automotive engineers have been involved in the development of forward looking systems that help the driver to recognize detectable threats in his driving path. One of the first cars that was equipped with a radar based obstacle detection system was built by the Delco Radio Division of GM Corp. (Fig. 2).

Due to the tremendous progress in the field of microelectronics with respect to scaling, speed, complexity and costs, automotive suppliers are now in a position to provide millimetre-wave, laser, and vision-based obstacle detection systems. The most prominent examples are the radar or lidar-based adaptive cruise control systems that maintain either a constant vehicle cruising speed or a constant headway between vehicles.

Next generation systems will be able to handle low-speed as well. This is especially significant since a large percentage of accidents occur under urban dense-traffic conditions.

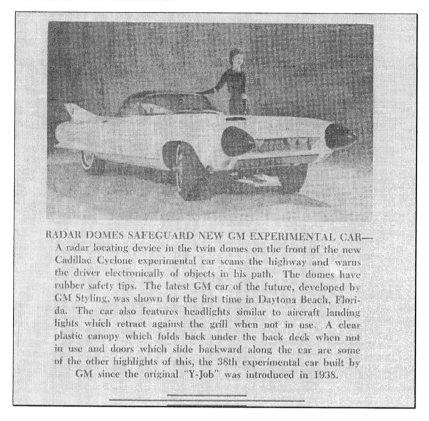

Fig. 2 Clipping from DELCO Radio Broadcaster, Vol. 23, No. 3, Kokomo (Indiana), March 1959

Another class of obstacle detection systems is represented by night vision systems which enhance the driver's vision at night time. Passive systems already on the market use a head-up display to project a thermal image (7-14 µm wavelength range) acquired by a thermal imaging device into the driver's field of view [3]. Active night vision systems operate in the near-infrared (NIR) wavelength range (780 – 1100 nm) and consist of a NIR-headlamp, a CCD or CMOS-imager based camera and a display that will be realized either as a HUD or an LCD display located in the instrument panel [4]. The approaches mainly differ in the way the scene is represented: active systems are able to provide a more "natural", higher-resolution image that includes lane markers, traffic signs, etc. while passive systems do not interfere with oncoming headlights and outperform with regards to range and resistance to adverse weather conditions.

Finally, it must also be noted that obstacle detection is not limited to exterior sensing. Basically, similar approaches used for short range sensing (i.e. low-power ultra-wide band radar, lidar and electro-optical scanning) can be used for interior sensing systems like occupant position sensing or theft protection [5].

The following sections introduce the most common technologies and conclude with a summary of Delphi's strategy to integrate obstacle detection systems into a comprehensive safety concept.

2 Technologies

Competing technologies for obstacle detection can be divided into three groups:

- **Radar:** This is an acronym for **ra**dio **d**etection **a**nd **r**anging and defines a device that transmits electromagnetic signals (f = 100 kHz ... 300 GHz) and receives echoes from objects (targets) within its volume of coverage. Information provided by a radar device includes distance by measurement of the elapsed time between transmission of the signal and reception of the echo, range rate by measurement of the Doppler shift, and direction by the use of directive antenna patterns.

- **Light wave technologies using active illumination sources (Lidar, laser-scanning):** Lidar (**l**ight **d**etection **a**nd **r**anging) based approaches measure the range by evaluating the time difference between the transmission of a light pulse and reception of the reflection caused by an object at a certain distance. The time difference can either be measured directly or by analysing the phase shift of an amplitude or frequency-modulated signal. The laser scan approaches are based on the point-wise illumination of an object using a laser beam and the evaluation of the reflection by a position sensitive (PSD) or imaging device (CCD, CMOS). Since the distance between only one point in space and the sensor can be computed, a mechanically or electronically scanned system must be used in order to acquire the range information in a scene.

- **Passive light wave technologies (Vision):** Vision based approaches focus on the analysis of single images or sequences that have been acquired by a monocular or stereo camera system. Vision sensors must be able to handle a wide range of illumination conditions (from bright light to tunnels, shadows, etc.), resistance against blur due to vehicle motion and blooming due to oncoming headlights and direct sunlight.

The different physical constraints caused by wavelengths and the characteristics of the selected sensing devices offer specific properties that are beneficial for certain applications. E.g. radar has the ability to determine range and range rate simultaneously and accurately. Monocular vision systems, in contrast, cannot provide range and range rate data, however, they are able to detect roadway markers, traffic signs, road features and scene objects. Lidar systems do not provide instantaneous measurement of velocity and must be eye-safe but are not subject to radio regulations as are radar devices.

Since obstacle detection systems used for collision avoidance or pre-crash sensing require robust object detection and classification capabilities to determine the kinematics and the impact of an imminent collision threat, multiple sensor and data fusion approaches can be applied to obtain a representative model of the vehicle's surroundings [6]. The mechanization of a typical collision warning / avoidance system that develops a model of the scene and issues a set of vehicle control commands depending on the desired system function is illustrated in Fig. 3.

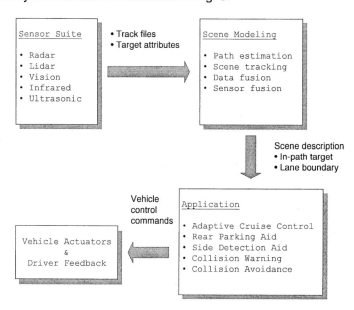

Fig. 3. Collision Avoidance System Mechanisation from [7]

3 Advanced Technologies

The interface between an obstacle detection system and the driver plays a crucial role in a collision warning system. If a warning is given to the driver the desired reaction must be to safely manoeuvre the vehicle away from the hazard. This requires the consideration of the following aspects:

- the warning must be designed to provoke a response without startling the driver.

- Depending on the physical fitness of the driver, the system's reaction time should be adaptable. E.g., if a longer reaction time is factored into the system for elderly drivers, then drivers with a youthful disposition might regard the warning as unnecessary and annoying.

- False alarms must be minimized. Otherwise the driver might begin ignoring the warnings.

Fig. 4. Exterior View of a Concept Car That Demonstrates Delphi's Vision of Integrated Safety Systems

To tackle this issue, Delphi has attempted to monitor the key driver characteristics in addition to the surrounding areas of the vehicle. The objective of this approach is to provide information that allows a workload management system to assess the driver's state in order to provide appropriately tailored warning cues. A concept car, shown in Fig. 4, has been built for the purpose of customer presentations. It includes a suite of exterior (76 GHz and 24 GHz long and short range sensors for forward, rear and side sensing, vision systems for lane tracking and anticipatory crash sensing) and interior sensors (respiration and heart rate monitor, stereo vision eye tracking for driver monitoring). It also includes a human-machine interface that helps the driver focus on potential threats. It emphasizes the effects of workload management on collision warning and infotainment systems while providing capabilities for "static" and "on-road" demonstrations [8].

4 Conclusions

The best way to protect vehicle occupants and to provide traffic safety is to avoid accidents. Obstacle detection systems are the key elements not only for active safety systems but also for occupant recognition sensors which help to enhance passive safety systems. As a result, the fusion of both approaches will drive future developments.

Acknowledgements

The author would like to thank the Advanced Vehicle Systems, Collision Avoidance and AED teams for their continuous support, and to thank Dr. Glenn R. Widmann for many valuable discussions.

References

[1] J.L. Blincoe; The Economic Cost of Motor Vehicle Crashes 1994, U.S. Department of Transportation (Report DOT HS 808-425), 1996.

[2] Bundesanstalt für Strassenwesen (BAST), "Volkswirtschaftliche Kosten der Personenschaeden im Strassenverkehr", Publication of BAST, Issue M102, January 1999, Authors Herbert Baum, K-J Hoehnscheid, University of Cologne.

[3] C. Buettner, "Thermal infrared for automotive applications", AMAA 2002.

[4] K. Eichhorn et al., "Verbesserte Nachtsicht mit Infrarot-Scheinwerfern", ATZ 9/2001, Jahrgang 103.

[5] S.-B. Park, "Optische Kfz-Innenraumueberwachung". Dissertationsschrift. Universitaet Duisburg. 2000.

[6] Volkswagen AG: Vorausschauende Kollisionserkennung, Bestellnr.: Z00.519.2 23.0.

[7] S.N. Rohr et al.: "An Integrated Approach to Automotive Safety Systems", SAE 2000 World Congress, Detroit (MI), SAE Paper 2000-01-0346.

[8] A. Teuner et al.: „A Strategy for Integrated Safety Systems", Salon d'Automobil 2001, Barcelona (Spain).

3.1.2 Environment Sensing for Advanced Driver Assistance - CARSENSE

J. Langheim

CARSENSE Consortium
Ave. du Technopôle, 29280 Plouzané, France
Phone: +33/298459443, Fax: +33/298495655
E-mail: jochen.langheim@autocruise.fr, URL: www.autocruise.net

Keywords: driver assistance, radar, laser, video, fusion

Abstract

The CARSENSE programme, grouping together 12 major industrial and research partners and sponsored by the EC, is to develop a sensor system, the purpose of which is to provide sufficient information on the car environment at low speeds to assist in low speed driving in complex (urban) environments. This article describes the main objectives of the programme, these being the improvement of the individual sensors and the merger of the information from these sensors in a fusion unit.

1 Introduction

After the introduction of Adaptive Cruise Control (ACC) into the market in 1999 all surveys and experimental assessments have shown a high interest and product acceptance for such systems. However, these first advanced driver assistance systems (ADAS) are very much limited to use on motorways or urban expressways without crossings. The traffic situations evaluated by these systems consist only of other vehicles, such as cars and trucks moving in simple patterns. Processing is restricted and can be focussed on few, well defined detected objects. Nevertheless, even for these relatively simple situations, these first systems cannot cope reliably with fixed obstacles. They also occasionally behave in an unexpected manner, causing surprise to the driver in 'cut-in' situations. Here, the width of the sensor beam may not fully cover the area in front of the vehicle, resulting in a late response from the ADAS.

2   CARSENSE Objectives

When such ADAS are in wider use, it will be necessary to extend the operation envelope to cover more complex situations, such as dense traffic environments in sub-urban or urban areas. Traffic is characterised by lower speeds, traffic jams, tight curves, traffic signs, crossings and weak traffic participants such as motorbikes or bicycles. Very soon road scenarios become very complex and it is more and more difficult to operate an ADAS reliably. This is because those sensor systems currently available for monitoring the driving environment provide only a very small amount of the information which is necessary to manage these higher level driving tasks. It has been identified in previous R&D programmes, that one of the crucial requirements for achieving significant progress in ADAS technology is a considerable increase in the performance of the driving environment monitoring systems. This includes larger range, greater precision and higher reliability of the sensor information, as well as additional attributes. The way to reach this goal is to improve existing sensors, such as radar, laser and video, as well as to fuse the information/output of these different sensor systems.

CARSENSE will develop a sensing system and an appropriate flexible architecture for driver assistance systems, with the aim being to advance the development of ADAS for complex traffic and driving situations, initially at low speeds. The ADASE project identified that driver assistance at low speed is the next most feasible function after the introduction of ACC. However, this functionality requires two of the fundamental steps towards future ADAS: reliable information about stationary objects and a wider field of view, albeit only in the near range.

Based on these agreed scenarios, chosen to cover a large spectrum of low speed real-life situations, data from various sensors will be acquired simultaneously and recorded along with additional driving sequence description scripts. The scripts provide the partners with the means for calibrating their sensors and for testing their algorithms. LCPC (LIVIC department) will use it's own test tracks to achieve this task.

At the end of the project, a second set of tests will permit a study of the performance of each newly designed sensor. These will also help assess the benefits of sensor fusion and allow an evaluation of the detection performance of the entire system, and its capacity for coping with low speed driving situations.

3 Datalogging and System Architecture

The CARSENSE system is a multi-sensor data fusion system designed to detect objects in front of the host car. This multi-sensor data fusion system consists of a set of internal and external sensors from where information is fused within a single data fusion unit. Internal sensors give information about the host vehicle state, such as its velocity and steering angle information. External sensors (Laser, Radar, and image sensors) sense information outside the vehicle, such as obstacles and road information. All the sensors and the data fusion unit are connected via CAN busses. A system specification of CAN messages has been built according to external sensors constraints [1]. [2] shows a similar approach.

- The development process, based on data collection and off-line processing, requires a reliable and powerful data logging equipment. The datalogger designated for this purpose has been developed by ENSMP. It is installed in the first architecture representing in addition an essential component for the control of the data flow within this architecture [10]. This CARSENSE system is implemented on a test vehicle (Fig. 1).

Fig. 1. Test Vehicle Alfa 156 Sportwagon 2.0 Selespeed

The following paragraphs outline some new hardware and software developments being done within the framework of the CARSENSE project.

4 External Sensors and Data Fusion Hardware

Three kinds of sensors are embedded in the system (radar, laser and video sensors). Each one will have an intelligent processing unit and a CAN interface. A specific hardware is being developed for data fusion.

4.1 Radar Sensor

Prototype radar sensors (Fig. 2) have been developed by Thales. They will later be commercialised by its subsidiary Autocruise. The technology used GaAs based Monolithic Microwave Integrated Circuits (MMIC). MMIC based radars have the potential to be produced at a low cost level that is acceptable to automotive customers [3].

Fig. 2. Autocruise Radar Sensor Sample

The main technical characteristics of the radar are shown in Table 1.

Frequency:	76-77 GHz
Range:	<1 - > 150 m
Search Area:	11°
Speed meas. precision:	< 0.2 km/h
Angular Precision:	< 0.3°

Table 1. Main Technical Characteristics of the Radar

At the core of the radar is the transmitter / receiver module (T/R – Module). It contains all the microwave circuits necessary for the radar function.

Within the CARSENSE project, this radar sensor will be improved by Thales on two fronts:

- Widened field of view in the short range area (+/- 35 up to about 40 m) by use of a new optical part for the radar antenna
- Improved detection of fixed targets by use of a new waveform

In order to improve the later one, a Digital FM radar waveform will be used that combines the advantages of frequency shift keying (FSK) with the advantages of Frequency Modulation (FMCW) .

This waveform has the following advantages:
- Allows a very high speed discrimination
- Allows fixed object detection and discrimination
- Ideal compromise for highway /road operations
- No distance nor speed ambiguity
- Low sensitivity to interference

4.2 Laser Sensor

The New IBEO laser-scanner LD ML Automotive (Fig. 3) is a high resolution scanner with an integrated DSP for sensor-internal signal processing. The laser-scanner emits pulses of near infrared light and measures the incoming reflections of those pulses. The distance to the target is directly proportional to the time between transmission and reception of the pulse. The scanning of the measurement beam is achieved via a rotating prism. The measurements of one scan form a 2.5-D profile of the environment.

Fig. 3. New IBEO Laser scanner LD ML Automotive

Expected characteristics (Prototype of the new laser sensor) are shown in Table 2.

Viewing Angle:	up to 270°
Angle Resolution:	0,35°
Scan Frequency:	10 Hz
Distance Measurements Reflecting Targets:	up to 150 m
Dark Targets (reflectivity 5 %):	up to approx. 30 m
Distance Accuracy:	+/- 5 cm (1 sigma)
Eye-safe:	laser class 1

Table 2. Expected Characteristics of the New Laser Sensor

3.1.2 Environment Sensing for Advanced Driver Assistance – CARSENSE

4.3 High Dynamic Range Video System

A video sensing system is being developed for sensing features such as stationary and moving obstacles and road information. This will consist of multiple cameras and a dedicated embedded processing unit. The cameras are being developed by Jena-Optronik GmbH, exhibiting a high dynamic range (to cope with ambient lighting conditions outside the vehicle) and a high resolution. The purpose of the processing unit is to enable image processing algorithms to operate at real time (frame rate ~ 25 Hz), while demonstrating a clear relationship between the development hardware and that which will go into production. (Fig. 4)

Fig. 4. High Dynamic Range Video System

TRW Automotive is developing a high performance platform (Fig. 5) which contains both a Field Programmable Gate Array (FPGA) and Digital Signal Processor (DSP) which is capable of processing raw video at a high data rate. The hardware will consist of an embedded micro controller to handle communication with the rest of the subsystems.

The unit is modular and stackable, so that multiple algorithms can be tested on additional processing boards, and those algorithms that require more processing power than is provided by a single FPGA/DSP subsystem can make use of multiple processors, if necessary.

This development platform will demonstrate that complicated image processing algorithms can be realised using a cost effective embedded hardware solution.

4.4 Data Fusion Hardware

The requirement for the architecture to be flexible has lead to an Object-Oriented approach being taken. This leads to an open hardware requirement, with the software design (developed by TRW) being platform independent. The choice of platform is driven by the requirement for enough processing power for the data fusion algorithms.

Fig. 5. Embedded Image Processing & Fusion Unit

Initially algorithm development will make use of a PC platform, for ease of implementation. The final solution will be implemented on an embedded platform.

5 Algorithm Developments

Video processing and data fusion algorithms will be developed by INRIA, LCPC (video and data fusion), INRETS-LEOST, and TRW (data fusion).

5.1 Video Processing

The vision task (LCPC) can be split into two main topics : Lane Marking Detection and Obstacle Detection.

Lane Marking Detection

The robust detection and tracking of lane markings and lane boundaries, using on-board cameras, is of major importance to the CARSENSE project. Lane boundary detection can assist the external sensors in identifying whether obstacles are within (or out of) the host vehicle's lane. Road boundary detection must at least provide, with high accuracy, estimates of the relative orientation and of the lateral position of the vehicle with respect to the road (Fig. 6).

Two approaches, based on different road model complexity, will be tested:

First, a real-time algorithm [6] will allow computation of the orientation and lateral pose of a vehicle with respect to the observed road. This approach provides robust measures when lane-markings are dashed, partially missing, or perturbed by shadows, highlights, other vehicles or noise.

The second approach [7], contrary to usual approaches, is based on an efficient curve detector, which can automatically handle occlusion caused by vehicles, signs, light spots, shadows, or low image contrast.

3.1.2 Environment Sensing for Advanced Driver Assistance – CARSENSE

Fig. 6. Lane Marking Detection

Obstacle Detection Through Stereovision and Fusion

Here, the aim is to detect obstacles located at less than 50 m in front of the test vehicle. For CARSENSE, an obstacle is a vehicle (car, truck), a motor bike, a bicycle or a pedestrian cutting into the host vehicle's trajectory.

A binocular vision and multi-sensor fusion approach is proposed, which allows the detection and location of such objects. Thus, the matching of data from the two cameras makes it possible, via triangulation, to detect objects located above the roadway and to locate them relative to the host vehicle. The matching process may use results obtained from other types of sensors (rangefinders) in order to make reliable detection and increase the computational speed (co-operative approach).

5.2 Data Fusion Processing

Multiple sensors, apart from providing different coverage, can detect the same object, but with differing accuracy of the parameters describing that object (for instance, range and angle). This information is complimentary, and leads to a measurement of higher integrity, accuracy and confidence. In addition to this, certain sensors may see information 'invisible' to other sensors, such as video's ability to locate road markings. This assists in improving the positioning of objects with respect to the real-world, rather than the subject vehicle.

Data Fusion involves, as its name implies, the merging of the information provided by different sensors in order to get a better picture of the environment in front of the vehicle. The different sensor units deliver processed information about the road geometry (curvature, etc.). They also deliver information about relevant objects detected in the vicinity of the CARSENSE vehicle, in the form of a list of objects. Each object is characterised by a certain number of attributes, such as position, velocity, etc., the quality of which depends on the sensor / processing unit under consideration. The goal is to develop vision algorithms in order to detect obstacles on the road and to produce the trajectories of the various object of the scene (other vehicles as well as static obstacles). To achieve this goal INRIA will develop algorithms based on motion analysis, that enable to compute the dominant image motion component, assumed to be due to the car motion. The principle of this algorithm is to determine the polynomial model (constant, affine, and quadratic), which closest describes the image motion in a specified zone of the image by statistical multi-resolution techniques. Detection of obstacles can then be done and trajectories of these objects, as well as the time to collision, can then be computed and provided as input to the car system.

6 Summary and Conclusion

The CARSENSE project is an important step on the way to high performance perception systems in future ADAS. The combination of multiple sensor information will improve object detection reliability and accuracy over that derived from today's sensors. With the development of new sensor functions, such as detection of fixed obstacles and wider field of view, the systems will be capable of use in urban areas, and high integrity (and comfort) ACC systems. Ultimately these sensing systems may be used in safety applications such as Collision Mitigation or even Collision Avoidance.

Acknowledgments

The authors would like to thank the European Commission for its support to make this programme possible.

Background Information

Research Partners

Autocruise, BMW, CRF, Renault, IBEO, Jena-Optronik, Thales Airborne Systems, TRW Automotive, INRIA, INRETS-LEOST, ENSMP, LCPC-LIVIC.

References

[1] Martine Wahl: "CARSENSE deliverable D4.a-ARCHITECTURE", CARSENSE European project (IST-1999-12224), Project Restrictive Public, 30 pages, September 26, 2000.

[2] Martine Wahl, Guillaume Doree: "A CAN application layer for an experimental real time obstacle detection study", IEEE International Conference on Intelligent Transportation Systems (ITSC'2000), MI, USA, p.276-281, October 1-3, 2000.

[3] Langheim, J., Henrio, J.-F., Liabeuf, B. : ACC Radar System « Autocruise » with 77 GHz MMIC Radar. ATA. Florence, 1999.

[4] Kirchner, A.; Lages, U.; Timm, K.: Speed Estimation with a Laser Rangefinder. Proceedings of ICARCV'96: 4th International Conference on Control, Automation, Robotics and Vision, 04.-06.12.1996, Singapore, pp. 1875 - 1879.

[5] Lages, U.: Umgebungserkennung mit Laserscannern. Tagung zur "Abstandsregelung (ACC)", 08. – 09.12.1999, Essen.

[6] Jean-Philippe Tarel, Frédéric Guichard, and Didier Aubert: "Tracking Occluded Lane-Markings for Lateral Vehicle Guidance", IEEE International Conference Circuits, Systems, Communications and Computers (CSCC'99), July 4-8, 1999, Athens, Greece, p 154-159.

[7] Jean-Philippe Tarel and Frédéric Guichard, "Dynamic Tracking of Lane Markings and Boundaries in Curved Road", IEEE International Conference on Image Processing (ICIP'2000), September 10-13, 2000, Vancouver, Canada.

[8] Langheim et al.: "CARSENSE –Sensing of Car Environment at low speed driving", ITS Turin, Nov. 2000

[9] F. Nashashibi et al.: "RT-MAPS: a framework for prototyping automotive multi-sensor applications". Ecole des Mines de Paris (France). IEEE Intelligent Vehicles Symposium 2000. Dearborn, MI, USA, October 3-5, 2000.

[10] Thompson, M.J., Trace, A.L., and Buchanan, A.J., (2001). "An Embedded Image Processing System for Driver Assistance System Algorithm Development", 4th International Conference on Vehicle Electronic Systems, Coventry, UK.

[11] Shooter, C., and Buchanan, A.J., (2001). "A Flexible Architecture for the Development of Multi-Sensor Driver Assistance Systems", 4th International Conference on Vehicle Electronic Systems, Coventry, UK.

3.1.3 True 360° Sensing Using Multiple Systems

T. Goernig

Conti Temic microelectronic GmbH
Ringlerstrasse 17, 85057 Ingolstadt, Germany
Phone: +49/841/881-2493, Fax: +49/841/881-2420
E-mail: thomas.goernig@temic.com, URL: www.temic.com

Keywords: 360° sensing, PMD, pre crash sensing, safety, integration, IBA

Abstract

Today's automotive passenger restraint systems see a major change from passive safety systems towards active safety systems. The development of passive safety systems is focused on an improved performance of the restraint devices while the need for the protection of all road traffic participants is increasing. This requires new generations of sensors like occupant detection sensors and pre crash sensors, placed on dedicated positions of the vehicle to merge passive safety and active safety.

1   Development of Vehicle Safety Systems

Since the late 80's automotive safety systems have become standard equipment for most vehicles that we can find on the road today. These systems are well known as airbag systems and are state of the art of passive safety systems. Improved protection devices like foot bags, knee bags, inflatable belts and active headrest are under development. All these systems have in common that the decision to activate the restraint devices such as pretensioners, front airbags, side or head airbag is based on the signals from the acceleration sensors. The accelerometers are installed in the airbag control unit and also in the side and upfront sensors that are used to early sense the impact.

The development of the restraint devices is following the general trend to more individually controlled devices as it can be seen on airbag modules. First systems used single stage inflators, the next generation was equipped with dual stage devices and newer developments are based on multi stage inflator technologies such as variable output inflators (VOI). In the future we will also see more sensors as well as more sensor networks in automotive vehicles than today.

2 Sensing Requirements

With the introduction of the different types of airbag systems also different type of crash sensors are required. For the first airbag systems with driver only or driver and passenger airbags a central crash sensing was required. The electronic modules where equipped with a single accelerometer or in cases where also an offset performance was required, two accelerometers in a ± 45° orientation relative to the x-axis have been installed. This is the most common arrangement still used in today's airbag systems. With the introduction of thorax side airbags and later on head protection systems like Inflatable Curtain (IC) or Inflatable Tubular Structure (ITS), additional side sensors where required. In the first side sensors an accelerometer and a micro controller was used for signal pre processing. The signal was a PWM signal equivalent to the severity of the impact. Today's side sensor elements have a digital communication, transmitting the actual crash signal to the airbag control unit, where the signal is processed in the crash algorithm.

Upfront sensors represent the latest level of innovation of a passive restraint system. They are installed close to the front of a vehicle and are used to detect a crash signal in the earliest possible phase of the impact. The purpose is a better discrimination of certain crash scenarios as well as to calculate the severeness of the crash to activate a smart airbag system accordingly.

For some vehicle types such as off road vehicles or SUV´s, roll over sensors are currently introduced. The necessary components, gyros and low-g z-sensors are integrated directly into the airbag control module.

Future sensor systems will monitor the occupant and enable an adaptive restraint performance, taking in account the physical condition of the occupant and the real demand from the actual crash situation.

3 Sensor Technologies

Early airbag systems used mechanical crash sensors like the Ball and Tube sensor from Breed or the Rollamite from TRW. These sensors where mounted on structural elements in the front of the vehicle. The electrical contacts of this sensor type were closed on an impact and the airbags are inflated. A discrimination of the crash pulse was not possible with this type of sensor. The next big step in the development of airbag control modules was the introduction of so called single point sensing systems. This type of systems became possible with the introduction of electronic crash sensors. The early generations have been piezoelectric sensors, followed by bulk micromachined resistive sensors and surface micromachined capacitive sensors. An improved discrimination of the crash pulse became possible with the development and introduction of this type of sensors. Future airbag control modules will use sensor microsystems (MEMS/MST) [1], which are in the development in various research projects.

4 From Passive Safety to Active Safety

When we look at the future of the safety systems that we know today, the so-called passive safety systems, we have to realize that the penetration of passive airbag systems in the market has reached a relative saturation. The major potential can be seen in the further development of smart airbag systems. Also a shift from passive safety towards active safety ban be seen on the market. This is obvious with the increased number of systems that are used for improved braking like ABS, BAS or vehicle dynamics control systems like ESP and roll over prevention systems. A further development towards active safety systems are advanced ACC systems, Stop & Roll and Stop & Go systems.

Future active safety system applications are e.g. lane detection, lane keeping support, driver warning and driver assistance systems, blind spot detection systems, crash avoidance and automatic cruising systems, which me might see in the very long future.

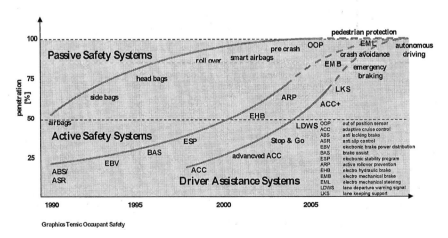

Fig. 1. Market Trends

5 Requirements for Future Safety Systems

From the actual road traffic and accident statistics it is known that passive safety systems have proven be very efficient in the reduction of injuries in vehicle crashes. From the same statistics it is also known that there are cases where it is necessary to improve the performance of a passive safety systems. Statistical data also shows that protection for other road traffic participants is required. There is a large group of other road traffic participants like motorbike drivers or bike riders as well as a large group of pedestrians. This can also be noticed in the planned regulations regarding passenger protection systems of the European Union [2] and the self-commitment of the European Automobile Manufacturer Association ACEA [3], which is also a major step towards active safety systems in the near future.

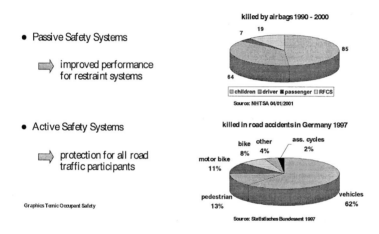

Fig. 2.   Statistical Information

6 Future Sensing System

Passive safety systems are activated in the moment when the impact occurs. The purpose of passive safety systems is to offer the best possible protection against injuries for the vehicle occupants, to limit the physical stress the occupant is subjected to. Due to this limitation a new generation of sensor and sensing systems becomes necessary to fulfil the requirements of active safety systems. For an active safety system it is required to get information about a possible impact before the actual impact occurs. Depending on the range of the sensor, the information is used for driver warning, for driver assistance and finally for crash avoidance. The signals will also be used for activation of a restraint system with a much higher precision and a better restraint performance compared to today's systems. For the improvement of the performance of the occupant protection system sophisticated interior sensing systems are required. The interior sensing system has to cover the driver and front passenger and in a second step also the passengers on the rear seats.

To generate reliable information before a possible impact occurs requires sensors that measure the proximity of objects within the surroundings of a vehicle. A true 360° sensing is required that consists of a front wide and narrow range sensing systems as well as a side and rear narrow range sensing systems.

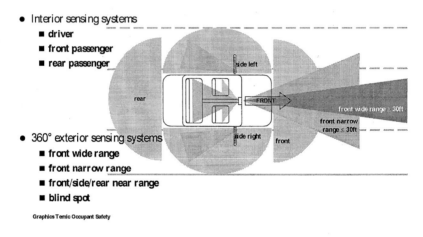

Fig. 3. Future Sensing Systems

7 Optical Pre Crash Sensor – Closing Velocity Sensor (CV Sensor)

A first generation of a proximity sensor consists of an optical sensor that measures the distance, velocity and direction of an object within the driving direction of the vehicle. With this information it is very early known before the impact occurs, which restraint device must be activated. It is also possible to recognize which severeness the impact will have. This information is used to control restraint devices like adaptive force limiters or adaptive airbag systems as well as reversible restraint devices like reversible grabbers and reversible pretensioners.

The CV sensor is mounted behind the windscreen, in a central position close to the rear mirror. The sensors principle is based on the speed of light. Infra red coded laser light is sent out and reflected at objects in the proximity of the sensor. The decoder circuit calculates the time of flight between the emitted and received laser pulse. A multiple beam arrangement has been selected for the sensor system to cover the complete area in front of the vehicle. With this multiple beam arrangement it also possible to calculate the direction of the approaching object. The speed of an approaching vehicle is calculated from the distance information of repetitive measurements. Besides the primary function of the proximity sensor, secondary functions such as rain sensor or light sensor are also possible.

A further target is to use the CV sensor as a sensor for a first generation of pedestrian protection system, as they have to be introduced in Europe from mid 2005 on.

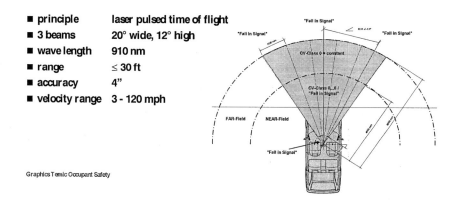

- principle   laser pulsed time of flight
- 3 beams   20° wide, 12° high
- wave length   910 nm
- range   ≤ 30 ft
- accuracy   4"
- velocity range   3 - 120 mph

Graphics Temic Occupant Safety

Fig. 4. Pre Crash Sensor

8 Optical Interior Sensor - Photonic Mixing Device (PMD)

The next generation of sensor that is targeted to be used for active safety applications is also an optical sensor. This sensor is based on the photonic mixing device technology [4]. With this technology it is possible to get a two-dimensional grey image and a distance information from each pixel of the image. With this type of optical sensor will enable object detection, identification and classification. The sensor technology can be used for interior and exterior vehicle applications.

In an installation in the passenger compartment the primary function of this sensor is passenger detection and classification as an advanced OOP sensor [5] to fulfil the legal U.S. requirements as defined in the revised

FMVSS208 [6], advanced occupant detection. The advanced image processing will also allow detection of child or infant seats in the passenger seat. As secondary functions head rest positioning, automatic seat and mirror adjustment based on the position of the head of the driver are possible. Another secondary function would be the use of an anti theft system.

- emission of RF modulated incoherent light
- reflection on desired 3D object
- detection of reflected light
- mixing with originally emitted signal on chip
- grey scaled image
- distance information from each pixel
- 3D image data advantages
  - high signal resolution
  - easy data processing
  - random pixel access

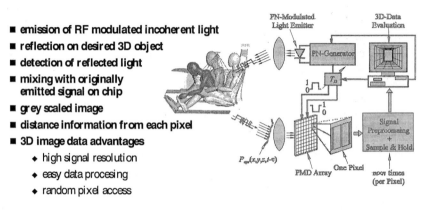

Graphics Temic Occupant Safety

Fig. 5. PMD Sensor

9 Exterior Sensing Systems

As an innovative sensor for active safety applications the PMD sensor system would also be installed behind the windscreen. The sensor is targeted to be used as an advanced pre crash sensor with an improved discrimination capability. With this capability it will be possible to analyse complex traffic situations and reliably activate driver-warning systems as well as to supply the necessary information for driver assistant systems or crash avoidance systems. With this sensor it will also be possible to identify objects in front of the vehicle, e.g. pedestrians, to activate the required protection systems.

A further sensor for active safety applications is the radar based sensing system. This sensor is installed behind the bumper, fan or front light area. The sensor is used today for adaptive cruise control systems for driver assistance. A next generation is an IR based high resolution scanning system that is also used for driver assistance application like ACC, Stop & Go or City ACC (Stop & Roll). This type of sensor can also be used in active safety applications like collision warning, emergency braking and collision avoidance.

A different type of sensor is used as a driver assistant system on the sides of the vehicle. A combined sensor system that is installed in the right and left side mirrors consists of an IR sensor and a camera system. It is used for lane detection, lane-keeping support and together with a backup long range radar as blind spot warning or lane change support system.

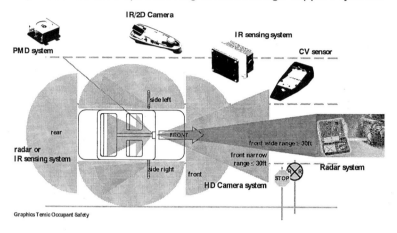

Fig. 6. Exterior Sensing Systems

10 Active/Passive Safety – System Integration

The trend towards active/passive safety puts new demands to the design of restraint systems. Due to the increased number of sensors the neccessarity to build up sensor clusters is obvious. Another important issue is the wiring effort for this type of systems. It has to be reduced for costs, assembly, weight, quality and reliability reasons.

Part of future safety systems will also be sensor clusters that include all the sensors, which are necessary to provide an optimum on passenger protection as well as to supply all relevant information from this sensor elements to other control modules in the vehicle.

Bus systems will be used for the connection and transmission of information between the different modules. There will be different types of bus systems, depending on the different applications that we can find in future vehicles, there will be passive safety system related bus systems like the Bosch/Siemens/Temic Sensor and Deployment Bus [7][8] and the related development of bus compatible igniters for the actuators of the passive safety system like it is done in the cooperation project IBA [9][10].

Further on there will be bus systems for x by wire applications and vehicle bus systems for general-purpose applications.

The different modules in the vehicle will also supply their own status information to the active/passive safety control module or other modules for an optimum use of information and the best possible functions for the safety on the road.

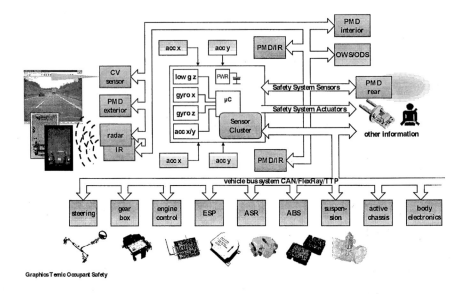

Fig. 7. System Integration

11 Conclusion

As it can be seen, there are many different requirements for the sensors and also different technologies for the sensors itself. There are also different sensing ranges, resolutions, wide and narrow focused sensors needed to fulfil the application of a vehicle surrounding sensing system. There will not be a single sensor for a true 360° sensing system in the near future, as the paper shows, using multiple sensor systems is a good approach of merging active and passive safety [11], to increase safety for all road traffic participants. The challenge for the future is to find the best possible combination of the different sensors and technologies to get a maximum of performance at economical costs.

Acknowledgements

The author like to thank the involved partners of the IBA project: G. Kordel (Dynamit Nobel GmbH, Fuerth), W. Schmid (NICO Pyrotechnik Hanns-Juergen Diederichs GmbH & Co. KG, Trittau), Dr. B. Loebner (Technical University Chemnitz, Department ET/IT, Center for Microtechnologies), Dr. H. Laucht (TRW - Airbag Systems GmbH & Co. KG, Aschau/Inn). As well as A.D.C. Automotive Distance Control Systems, Lindau and Conti Temic Sensor Systems, Kirchheim for supporting this paper with product pictures.

References

[1] S. Krueger, W. Gessner, AMAA Compendiums 2000 and 2001, www.amaa.de

[2] European Commission, Proposal for a European Parliament and Council Directive relating to the protection of pedestrians and other road users in the event of a collision with a motor vehicle, III/5021/96 EN, Buessels, 1996.

[3] COM(2001)389, July 11th 2001, ACEA Commitment relating to the protection of pedestrians and cyclists, europa.eu.int

[4] R. Schwarte, B. Buxbaum, H. Heinol, Z. Xu, J. Schulte, H. Riedel, P. Steiner, M. Scherer, B. Schneider, T. Ringbeck, New Powerful Sensory Tool in Automotive Safety Systems Based on PMD-Technology, AMAA 2000, Berlin, Germany, April 6-7, 2000, www.amaa.de

[5] Dr. H. Riedel, Paper 2002-01-1306, 3D Vision Systems for Active Safety, SAE2002 World Congress, Detroit, Michigan, U.S.A., March 4-7 2002, www.sae.org

[6] Title 49 Transportation; Chapter V - NHTSA, DOT, Part 571 FMVSS, §571.208 Standard No. 208; Occupant Protection, www.access.gpo.gov/nara/cfr/waisidx_00/49cfr571_00.html

[7] K. Balzer, C. Zelger, T. Goernig, BST Deployment and Sensor Bus, Airbag 2000+, 5th International Symposium on Sophisticated Car Occupant Safety Systems, Karlsruhe, Germany, Dec. 4-6, 2000.

[8] Common Bosch Siemens Temic Bus Description Rev.: 2.0, Apr. 24th 2000, www.temic.com

[9] G. Kordel, W. Schmid, Dr. B. Loebner, Dr. H. Laucht, T. Goernig, IBA Status Seminar, Ingolstadt, Germany, Nov. 27th 2000.

[10] T. Goernig, Integrated Electronics for Bus System Igniters, AMAA 2001, Berlin, Germany, May 21-22, 2001, www.amaa.de

[11] T. Goernig, Innovative Sicherheitssysteme im Kfz, Frankfurt, Germany, July 4-5, 2001.

3.1.4 Electronic Scanning Antenna for Autonomous Cruise Control Applications

B. Kumar[1], H.-O. Ruoß[2]

[1]BAE SYSTEMS Advanced Technology Centre, Elstree Business Centre
Elstree Way, Borehamwood, Hertfordshire WD6 1RX
Tel: +44/20/8624-6586, Fax: +44/20/8624-6099
E-mail: bal.kumar@baesystems.com

[2]Robert Bosch GmbH
FV/FLO, P.O.Box 10 60 50, 70049 Stuttgart, Germany
Tel: +49/711/811-7698, Fax: +49/711/811-6301
E-mail: oliver.ruoss@de.bosch.com

Keywords: electronic beam scanning, faraday rotation, ferrite, millimetric wave lengths, adaptive cruise control

Abstract

The successful integration of an electronic scanning antenna within car radar will allow the angular position of vehicles to be determined accurately and tracking to be maintained as the vehicle changes lanes and manoeuvres around corners. This paper describes the work currently undertaken at the BAE SYSTEMS Advanced Technology Centre to develop systems based on ferrite devices that are capable of electronically steering circularly polarised microwave beams, in the azimuth plane.

1 Introduction

Electronic beam steering antennas use non-mechanical components to alter the direction of a microwave beam and, by altering the gradient of the phase taper across the radiating aperture, control the angle of deflection. BAE SYSTEMS Advanced Technology Centre and Bosch GmbH have developed a quasi-optical ferrite device that is capable of steering a circularly polarised 77 GHz beam electronically. The device uses the magnetic gradient in the ferrite material to deflect the beam.

The successful integration of such a device within the car radar offers several advantages over the present three-beam radar system that relies upon amplitude comparison from adjacent receivers to determine the angular position of vehicles. Electronic scanning offers the ability to move

3.1.4 Electronic Scanning Antenna for Autonomous Cruise Control Applications

a single radar beam in extremely small increments to cover the required angular scan rapidly. This allows the position and the shape of the vehicles to be determined with greater accuracy and tracking to be maintained as the vehicle changes lanes and manoeuvres around corners.

2 Theory

The scanning device relies upon Faraday rotation for its operation - i.e. when a plane wave is travelling through a ferrite material, which is subjected to a longitudinal magnetic field, its plane of polarisation rotates. The direction of rotation is independent of the direction of propagation and only depends upon the direction of magnetic field. This effect forms the basis of the scanning device. Biasing coils, which produce the magnetic field, are arranged such that each half of the ferrite is magnetised in opposing directions and there is a taper in magnetisation across the radiating aperture as shown in Fig. 1.

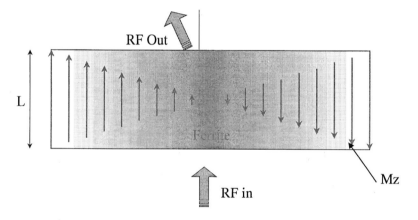

Fig. 1. Device Operation Principle

When a circularly polarised electromagnetic wave interacts with a block of ferrite magnetised as above, the wave experiences a differential phase shift across its aperture (i.e. relative to the centre there will be a positive phase taper on one side and negative on the other) resulting in a deflection of the beam. The angle of deflection can be controlled by the direction and the magnitude of the current flowing in the biasing coils.

3 Experimental Evaluation of the Ferrite Cell

The current 3-beam radar has an aperture of approximately 75 mm, however it is envisaged that the future radar system requirement will be focussed around a much smaller aperture size. The cell aperture of 50 mm has been selected to meet these requirements. The ferrite cell is constructed from commercially available material and the quarter wave plates are used to minimise the reflection from its surface shown in Fig. 2. The evaluation of the cell has been carried out using an HP 8530 W-band vector analyser, and the measurements have been repeated in an anochrobic chamber in Germany.

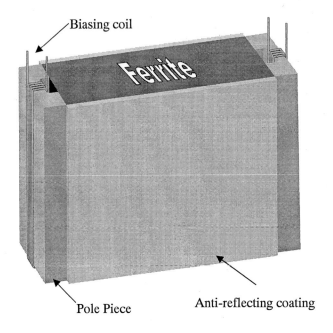

Fig. 2. Ferrite Cell

A corrugated feed horn launches linearly polarised millimetre wave radiation into a quasi-optical feed. The signal is then focused onto a plastic quarter wave plate and emerges from the feed and into free space circularly polarised. The received signal passes through the ferrite cell mounted on a turn-table (to allow the deflection angle to be measured) and a receiving lens behind the cell focuses the signal onto a receiving horn. A second quarter wave plate was inserted between the lens and the receiving horn to convert the received, circularly polarised, signal into a linearly polarised plane wave for optimum coupling into the receiver horn,

3.1.4 Electronic Scanning Antenna for Autonomous Cruise Control Applications

as shown in Fig. 3. Fig. 4 shows plots of the beam being scanned in the positive and negative directions.

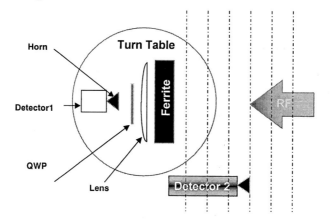

Fig. 3. Experimental Set-up

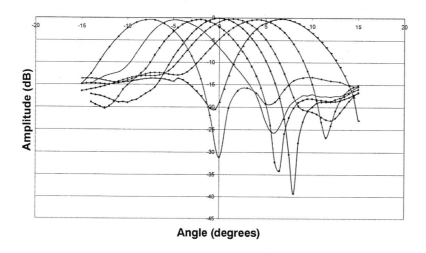

Fig. 4. Polar Plots as Function of Bias

4 Conclusion

We have demonstrated the principle of electronic beam steering using ferrite devices at 77 GHz in a laboratory environment. The integration of electronic beam scanning into automobile radar offers improved performance, robustness and size reduction amongst other benefits. It is envisaged that, in future, microsystems-based components (RF-MEMS) will further reduce the size and cost of such devices. In addition the RF power required for a single beam radar system can be significantly lower than that required by the three beam system. This reduction in power offers a number of advantages. Firstly, it allows the use of low power inexpensive Gunn diodes for the millimetric source. Secondly, the number of balanced mixers can be reduced from three to one, thus reducing the number of beam lead diodes from seven to three. Finally, the size of the MIC can be significantly reduced and components such as the Wilkinson power divider and the associated surface mount resistors can be removed.

5   Acknowledgement

The authors would like express their thanks to colleagues at the BAE SYSTEMS Advanced Technology Centre and Robert Bosch for their inputs and contributions.

3.1.5 Advanced Uncooled Infrared System Electronics

H. W. Neal, C. Buettner

Raytheon Systems Company
Dallas, TX 75265, USA

Keywords: uncooled infrared sensors, ferroelectric, non-uniformity correction, night vision, video processing

Abstract

Over the past two decades, Raytheon Systems Company (RSC), formerly Texas Instruments Defence Systems & Electronics Group, developed a robust family of products based on a low-cost, hybrid ferroelectric (FE) uncooled focal-plane array (FPA) aimed at meeting the needs for thermal imaging products across both military and commercial markets. Over the years, RSC supplied uncooled infrared (IR) sensors for applications such as in combat vehicles, man-portable weaponry, personnel helmets, and installation security. Also, various commercial IR systems for use in automobiles, boats, law enforcement, hand-held applications, building/site security, and fire fighting have been developed. These products resulted in a high degree of success where cooled IR platforms are too bulky and costly, and other uncooled implementations are less reliable or lack significant cost advantage. Proof of this great success is found in the large price reductions, the unprecedented monthly production rates, and the wide diversity of products and customers realized in recent years. The ever-changing needs of these existing and potential customers continue to fuel the advancement of both the primary technologies and the production capabilities of uncooled IR systems at RSC. This paper will describe a development project intended to further advance the system electronics capabilities of future uncooled IR products.

1   Background

IR imaging systems have been used by the military sector for many years to provide operational capability at night. Operation Desert Storm graphically demonstrated that the ability to locate, track, and confirm the kill of tactical targets in extreme darkness is a major advantage. Unfortunately, traditional cooled IR systems are typically large and costly. As a result, their use is restricted to aircraft or heavy-armour platforms. In recent years, however, the military realized that light armour, support vehicles, and the individual soldier need equivalent night vision capabilities for maximum support during night time manoeuvres. It was this crucial need that resulted in a whole new generation of small, low-cost IR sensors. As a result of their reduced cost and ease of use, these sensors also fit well into the paramilitary and civilian sectors, supporting law enforcement, border patrol, automotive, installation security, and numerous other industrial applications [1].

RSC began developing uncooled IR technology in the 1970s, based on the temperature-dependent pyroelectric properties of the FE material, barium strontium titanate (BST). Today, various military and commercial IR systems are being produced and marketed based on this original technology. With the continued support from the U. S. Army Night Vision & Electronics Sensors Directorate (NVESD) and the Defence Advanced Research Projects Agency (DARPA), RSC has proven that uncooled technology is not only a viable IR technology, but is also good business. A state-of-the-art detector factory has been built with the ability to manufacture hundreds of FE detectors each month with average f/1 noise-equivalent temperature difference (NETD) of 70 millidegrees Kelvin (mK), with better than 99.8 percent operability. Also, system production lines are turning out several hundred finished sensors of various configurations each month.

At every phase of the business, the costs have been of crucial importance. Efforts to raise efficiency and throughput while maintaining or enhancing system-performance parameters consistently use much of the available resources. Many programs include goals of cost reduction as well as the required repackaging or performance enhancements to drive the products closer to production readiness. One such program was the Technology Reinvestment Project (TRP). This program, jointly funded by DARPA and RSC, and managed by the Department of Transportation (DOT), exploited military technology in developing commercial products. At its core, the program used a series of component development efforts, including low-cost IR optics, high-volume FE detector processing, and next-generation system electronics to demonstrate significant progress toward system cost reductions and performance improvements.

2   General System Overview

The greatest advantage of an RSC uncooled FE IR system is its simplicity. Fig. 1 shows the inside of an automotive IR sensor based upon that technology. The IR energy in the 7 to 14 micrometer (μm) wavelength band from the scene is collected by the imager, modulated by a smoothly rotating chopper, and focused onto the uncooled FE detector array. Each

3.1.5 Advanced Uncooled Infrared System Electronics

detector pixel converts the absorbed energy, or heat, into electrical signals for use by the system electronics. After simple processing, the video signal is available to be displayed on a standard monitor [2].

Several different IR imager configurations have been developed for RSC's uncooled systems, which vary the field of view (FOV), resolution, range, and sensitivity for a given detector performance. These systems generally use f/1, refractive, diamond-point-turned (DPT), germanium elements with horizontal FOV's ranging from 4.5 to 40 degrees. These lenses are typically optimised for peak performance, making them too costly for some applications. The TRP made tremendous improvements in optics by developing new low-cost materials and processing techniques for extremely high volume capability. These improvements allow RSC to produce IR optics at a fraction of the cost of previous imagers with adequate performance for most uncooled IR systems. Fig. 2 shows examples of the improvements made with IR optics by the TRP.

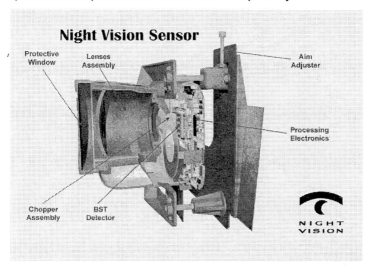

Fig. 1. The Inside of an Automotive IR Sensor

Because the FE detectors are AC-coupled devices as well as to ease the dynamic-range requirements of both the detector and electronics, RSC's uncooled systems use a chopper to modulate the IR energy of the scene. This chopper is a smoothly rotating disk that has an Archimedes spiral pattern to facilitate vertical sampling through the simple vertical motion of the edge of the spiral over the rows of the detector. It is driven by a long-life direct current (DC) motor and contains an optical sensor for speed and phase control. This disk can either be opaque or diffusing to IR energy. Opaque choppers, made of metal or plastic, allow the detector to reference the scene energy to the apparent temperature of the disk itself. Diffusing

TRP PROGRESS

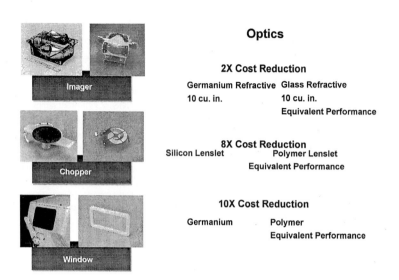

Fig. 2. The TRP IR Optics Progress

choppers, which have been made of germanium or silicon, reference the scene to a defocused image or to the local-scene average. In the opaque disk, the spiral pattern is cut out of the material. For the diffusing disk, the pattern is polished or embossed into the material, forming binary diffractive optic (BDO) lenslets. This BDO pattern creates small-area averages of the scene, which are used as the reference for the detector. The TRP advanced the technology of diffusing choppers in both the design of the BDO pattern and in the materials used. Several types of polymers have been studied in an effort to greatly lower costs, and the results are promising. Also, various motor types were investigated to find the best match for cost, size, and performance.

RSC's uncooled FE FPA development has evolved into today's detector, which places 320 x 240 pixels on 48.5 μm centres for a total of 76,800 active pixels in a 4:3 format. The detector consists of a reticulated array of BST FE ceramic capacitors bump-bonded to a readout integrated circuit (ROIC) that contains a preamplifier, noise filter, buffer and switch for each pixel, and a row-address shift register and column multiplexer for TV-compatible output. This hybrid array senses IR radiation from 7 to 14 micrometer wavelengths and boasts an NETD below 50 mK with f/1 optics. The device is packaged in a 40-pin, ceramic, dual-in-line package (DIP) with a germanium or silicon window, as shown in Fig. 3. It also contains a thermoelectric cooler (TEC) and a reference that are used for room-temperature stabilization. The TRP had a significant impact by developing high-volume production process techniques for the detector-manufacturing operation. Many processing steps, previously done sequentially on a single die, are now almost all done on a 150-millimeter wafer full of individual die. This has had a positive effect on performance, volume, and yield. As a result, RSC's FE detector factory leads the industry in its ability to produce large quantities of high-performance uncooled detectors at an extremely low cost.

Fig. 3. The Uncooled Ferroelectric Detector and Package

3 Electronics Overview

The overall strategy of this TRP electronics effort was to design and build a prototype system that would allow demonstrating and testing new circuit concepts intended to reduce the size, power, and cost of future products while enhancing their performance. This sensor maintained existing product optical and mechanical components and interfaces where appropriate to reduce the magnitude of the effort, and the existing electronics were replaced with a new implementation. Techniques were developed that maximized the level of integration without sacrificing flexibility. Many previously used components were eliminated. The circuitry was modularised and made expandable in an effort to broaden customer acceptance, facilitate an iterative design process, and improve product commonality.

After analysing the advantages and disadvantages of the existing products as well as a list of possible additional features, it was concluded that several aspects of the current electronics should be kept. The power conversion, TEC and chopper control, digital-to-analog conversion (DAC), and the TV output drive circuitry were duplicated in the new design by using portions of the custom analog application-specific integrated circuit (ASIC) and the entire power-conversion circuits from current products. The functionality of this portion of the circuitry is well understood and not described here. The remaining digital circuitry was replaced with a programmable component to adapt easily to changes. The device chosen for this task was the EPF10K100ARC-3 from Altera®. It is a 3.3-V, 100,000-gate complex programmable logic device (CPLD) in a 240-pin, surface-mount package. A functional block diagram of the TRP electronics is shown in Fig. 4. Many electronic functions were implemented in this single CPLD, including command and control of all modes and operations, timing generation, system configuration, serial communication, operator control interface, memory interface, and video processing.

4 Design Details

The command-and-control state machine section of the design responds to inputs from many sources and decides how to configure the system. The inputs come from external operator controls; the serial EIA/RS-232 port; a serial-configuration, electrically erasable, programmable, read-only memory (EEPROM); the system-timing state machine; and the video processing block. As a result of the various command inputs, the state machine generates the proper outputs for the other blocks to obtain the desired modes of operation. The three most fundamental modes are power-up, normal imaging, and gain normalization, but each of them has several possible configurations, depending on the inputs.

The timing generator creates all the clock signals that are required by the system. A 50-megaHertz (MHz) master clock, a power-up reset, and several signals from the command and control block are used to run the state machine. Depending on the status of these inputs, many of the outputs react differently. The state machine generates timing signals for detector clocks, memory access, serial port control, serial-configuration EEPROM download, video processing, and other miscellaneous signals. Most of these clocks typically run at specific frequencies that happen to be synchronized with RS-170 TV formats, such as 30-Hertz (Hz) frames, 60-Hz fields, 15.750-kilohertz (kHz) lines, 6.25-MHz pixels, and additionally 12.5-MHz memory and expansion port clocks. In certain modes, the memory clocks operate at 25 MHz. The clocks used to control the serial communications typically run at 9600 baud.

The system stores all configuration data in a serial EEPROM and downloads it on command or at power on, and is idle when not being accessed. The configuration data includes the default settings for things such as gain and level modes, video polarity, how the various video-processing blocks are set up, the gain-normalization algorithm settings, and many others. When the operator changes the initial configuration data, he can, through the serial communication port, reprogram the EEPROM with the new configuration. These settings become the new power-on defaults, until they are changed again. This feature will have a large impact

3.1.5 Advanced Uncooled Infrared System Electronics

on future products, since it will allow a common set of electronics to be configured differently without costly additional circuitry. Because this feature is integrated and programmable, configurations can be changed after the system leaves the factory.

The design also includes an integrated EIA/RS-232 serial communications port. This port allows remotely located operators or computers to have complete control of system functions or any of its configuration settings. The sensor constantly polls the port for a host controller, and establishes connection when one is present. It now runs at 9600 baud, but other speeds are possible. Also, the port handles all of its own handshaking and error checking to guarantee reliable operation. In order to verify the functionality of this port, the TRP also developed Microsoft® Windows NT™ and Windows 95™-based communications software, which proved quite useful in the development process.

The interface circuitry for the operator controls is used to convert switch inputs from the outside world to the proper commands for the video-processing block. The user controls are intended to be single-pole, normally open, push-button switches, with one contact tied to ground. This implementation will satisfy a simple, low-cost design in future products. The other contact is pulled to a high logic level on the board, so pressing the button will send a low logic level into the circuitry. The inputs are digitally de-bounced to prevent noise spikes, and have a time-delayed repeat capability, when needed. Some of the inputs, such as Gain-Up and Gain-Down, are used to drive a digital counter either up or down. Other inputs, like Gain-Mode, simply toggle a condition. At system power-on, the default condition is to accept inputs from the operator controls. At any time, however, the serial communications port can assume control and lockout the switches.

Besides the serial-configuration EEPROM, the system has two other memory devices, a 256 K x 16 static random access memory (SRAM) and a 256 K x 16 Flash memory. One-half of the SRAM contains the pixel gain-correction data for normal system operation, while the other half is used for the pixel offset-correction function. One-half of the Flash memory contains the power-on-default pixel gain-correction data, and the other half is reserved for a symbology screen. During the system power-up sequence, the symbology screen is read from the Flash memory and sent to the display. Next, the other half of the Flash is polled to determine if valid gain-correction data is present. A special code is stored in a certain address location, if data is present. If the Flash memory contains valid data, it is transferred to the SRAM for normal system operation. If that special code is not found, it is assumed that no data is present, and the SRAM is loaded with a unity gain-correction code. Once this sequence is complete, the Flash memory is disabled until needed again. During the gainnormalization process described below, the data in the SRAM for each pixel is iteratively modified. After this process is complete, the operator has the option of storing new default data into the Flash memory. This process simply copies the data from the SRAM to the Flash memory during video retrace time, and a busy indicator is displayed until it is complete. The use of the SRAM will be further explained in the video processing section below.

5 Video Processing Details

The TRP electronics project focused mainly on the video portion of the design, and many improvements were made. A functional block diagram of the video processing section is given in Fig. 5, and it will be explained in sequence. Some portions of the circuitry use external discrete components; some use sections of the custom analog ASIC; and the remainder are implemented inside the single CPLD. The video processing begins with the detector output signal. It is a serial stream of discrete voltage levels that represent a row-by-row readout of the uncooled array. The average DC level is about 2 V, and the maximum peak-to-peak voltage swing is also about 2 V. This waveform contains both the scene-dependent signal levels and the gain and offset nonuniformities of the array.

The detector output signal is first AC-coupled with a high-pass filter to remove any DC component of the waveform. The corner frequency of this filter is set very low (0.036 Hz), so that no field-rate droops are generated. Next, the signal is buffered with a video operational amplifier (Op Amp) configured as a voltage follower, so the gain is set to positive one. This Op Amp can also clip the output nicely if the input exceeds certain threshold voltages. It also has a very fast recovery time when it is returning from saturation, so it will not distort any adjacent pixels. This clipping capability of the Op Amp is quite important and protects the input stage of the next device from overvoltage. The video is then digitised with a 12-bit, video-speed, analog-to-digital (A/D) converter. This device has a 2 V input voltage range with an internally generated reference voltage and an input sample-and-hold circuit running at pixel rate, 6.25 MHz. The digital video is then sent into the CPLD for further processing.

Once inside the CPLD, many digital algorithms are performed to enhance the quality of the video. Each of these blocks can be individually controlled through a serial communications port. At the very input to the CPLD, a built-in-test (BIT) pattern can be selected instead of the incoming video, and two BIT patterns are possible: a checkerboard and a horizontal ramp. These patterns are specially designed to allow testing of all video-processing functions by a visual inspection of the output imagery.

3.1.5 Advanced Uncooled Infrared System Electronics

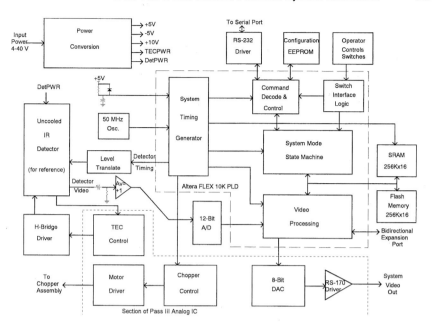

Fig. 4. The TRP Electronics Functional Block Diagram

The gain normalization stage is next in the chain. This block corrects for the gain nonuniformities of the uncooled array, and contains several components. As previously described, gain-correction codes are stored in the Flash memory and are downloaded into the SRAM at system power-on. During normal system operation, these codes are each sent to a 12 x 16 digital multiplier stage. The codes are 16-bits wide, and the video is 12-bits wide. Each of these gain coefficients represents a previously determined ratio of the array average value to the actual value of the individual pixel. Once the multiplication is performed for each pixel, the resulting output data stream will be gain-corrected. This algorithm can correct for nonuniformities in the range of one-half to two times the average, so the result of the multiplication has another resolution bit available. As a result, 13 bits of the output are selected, while the others are ignored. Besides the gain-correcting capability, this stage also eliminates defective pixels by substituting them with a previous good one. The gain coefficient for every pixel is also sent to decode logic to determine if it is the proper code. When the proper code is present, the substitution occurs by holding the previous pixel's value on the output of a latch for an additional pixel time. If appropriate, many pixels in a row can all be substituted with the value of a single good one, but no substitutions from a previous row of the array are allowed. The bad pixel in column number one cannot be substituted out, so it will remain bad in the output image.

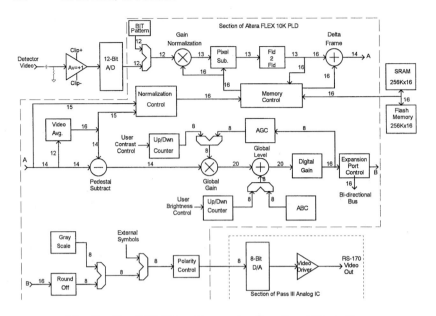

Fig. 5. The TRP Video Processing Functional Block Diagram

The process for obtaining the gain-coefficients and pixel-substitution codes is iterative. First, the system must be looking at a constant temperature scene, so each individual pixel's responsivity can be obtained. After the normalization button has been pressed, the SRAM is reset to a unity gain code for each pixel. Next, a low-pass filter, the same one that is used in the pedestal subtraction stage below, determines the array average and latches it in. The normalization control block compares the stored array average with the actual pixel value and determines if the gain is too low or too high. Then, each pixel's current gain code is read from the SRAM and is either incremented or decremented by one gain step. Once this is done, the new gain code is placed back into the SRAM to be used during the next field. This process continues for every pixel in the array. When it is time for the next field, the code that currently resides in the SRAM is not unity, but it has had some correction already performed on the pixel. Now this process is repeated until the operator is satisfied that all the pixels are sufficiently corrected. To further enhance the process, a turbo speed mode exists that allows the algorithm to change the gain code by more than one step. Various features, related to when and how often these turbo modes occur, are programmable. If the process continues for a pixel, and it determines that not enough range is available to properly correct the gain, a pixel-substitution code is placed in the SRAM for that pixel. Once a pixel has been replaced, it will not be restored until the process is started over again. Once satisfied, the operator can press another button to replace the power-up, default gain-coefficients in the Flash memory with the current

3.1.5 Advanced Uncooled Infrared System Electronics

values stored in the SRAM, making them the new defaults. This process takes less than 10 seconds to iterate to a good solution and about 30 seconds to store the data in the Flash. Future products will realize significant cost savings, since this is replacing a process that previously took about 1 hour.

The next major block in the video processing chain is the Delta Frame function. The purpose of this block is to remove the offset nonuniformities of the uncooled array. Rather than obtaining a one-time correction, as does the gain-normalization block, the offset correction is done continuously during system operation. To perform this function, it is necessary to exploit the fact that a pixel's voltage offsets are exactly the same in both polarity and magnitude in every field of output. The portion of the output video stream that corresponds to the scene information is also the same magnitude but is opposite in polarity in adjacent fields. As a result, the offsets are eliminated when adjacent fields of video are subtracted from one another. This is accomplished by first inverting every other field. Next, each field of video is simultaneously written into the SRAM and sent to a 13 x 13 digital adder, while the data from the previous field is retrieved from the SRAM and also sent to the adder. The result of this operation is that the adder continuously subtracts the previous field from the current field, a process called double-correlated sample (DCS) or Delta Frame [3]. Since the offsets are the same magnitude and polarity in both fields, they are eliminated. Also, the signal portion of the waveform is doubled, because it is the same magnitude but is opposite polarity in adjacent fields. The resulting doubled signal is also the same polarity in every field, because a field-to-field inversion was done before the subtraction. Since the technique is a DCS, it has the effect of raising the temporal noise by the factor of the square root of 2. As a result, there is an effective signal-to-noise ratio (SNR) improvement by the square root of 2. The digital adder output requires 1 additional bit of resolution, since the operation may have a carry out bit. The resulting video output now has both the gain and offset nonuniformities removed.

The digital video proceeds next to the pedestal subtract stage. This block removes large background-temperature offsets from the image. These offsets are generated as a result of the difference between the average scene temperature and the apparent temperature of the chopper as seen by the detector, and can be quite large with opaque choppers and can severely limit system dynamic range. When a diffusing chopper is used, this stage is relatively unnecessary, because the chopper uses the scene average as the reference and very little pedestal is generated. By definition, the pedestal is the average of the scene. To subtract it, a digital low-pass filter is used to obtain the scene average. This filter is an alpha-type filter that accepts 12 bits of the video and generates a result that has 16 bits of accuracy. This filter responds similarly to an analog RC-filter.

The scene average is obtained once each field and is subtracted from the video continuously during system operation with a 14 x 14-bit digital subtraction stage. The resulting image is essentially background-restored by this point in the processing.

Next is the global system-gain or contrast-control stage. This function is accomplished with an 8 x 14-bit digital multiplier, where the video is 14 bits wide and the gain control is 8 bits wide. This gain setting gives 256 possible gain settings, with a maximum gain of 1. The gain-control bits are either generated by an automatic gain control (AGC) section, or they are the counter outputs from the operator controls block. The gain is uniformly applied to the entire video every other field.

The AGC block samples the output video and decides if the dynamic range is optimised for the scene. If it is not set up properly, the block will attempt to adjust the global gain to correct. The total range of digital codes for the video contains four strategically positioned threshold levels, which represent minimum and maximum saturation points for both the white and the black ends of the range. The goal of the algorithm is to adjust the gain until the video fills the dynamic range just enough to allow saturation past the minimum points but not past the maximum points. The best gain setting is obtained when a predetermined number of pixels fall between the minimum and maximum thresholds at both ends of the video voltage range. The gain is allowed to adjust one step every other field, and it has provisions to prevent unnecessary oscillation between steps. It also has a fast speed, which will allow it to step faster when necessary to prevent delays in response time. When AGC mode is selected, the gain control stage uses the output of this block for its gain setting.

The next block in the video processing chain is the global level or brightness control. This stage adjusts the background level of the output image in order to obtain the most usable picture. This task is accomplished with an 8 x 20-bit digital adder, where the video is 20 bits wide and the level control is 8 bits wide. The width of this adder is necessary to prevent early clipping of the final output. As with the gain control, the adder selects from the output of the operator control counter and an automatic brightness control (ABC) stage. In this case, the ABC merely selects the appropriate centre for the video, depending on the system configuration settings. If the manual mode is selected, the operator can adjust the brightness through its full range. The desired global brightness is updated every field.

As a result of all the video processing described so far, it is now possible to improve the image by adding some gain by shifting the bits to the left until the desired gain is reached. The stage selects predetermined gain settings from 0.5 to 64, in multiples of 2. Each factor of two in gain reduces the output resolution by one bit. This block also clips the video at the

3.1.5 Advanced Uncooled Infrared System Electronics

appropriate points, depending on the gain settings to prevent erroneous data and to give the image a natural look. This function allows the system to have the desired dynamic range that was not possible with the raw detector video. At this point, the video is fully processed and ready for output, but there are a few blocks left, which add additional capability. A bi-directional expansion port exists to aid in testing and system-expansion options. This port is 16 bits wide and can either output existing video, input new or further processed video, or both. Timing is generated to allow these various modes to synchronize with the remaining system electronics. The overall purpose here is to allow video to be captured from the system. It also gives the capability for an expansion module to be designed that would provide additional capability beyond the standard processing, and then be input back into the system as though it were integrated with it.

Finally, the 16-bit video is rounded to the eight most significant bits to prepare to be converted back to analog and displayed. A 10-shade horizontal gray scale is added to the lower portion of the screen for testing and monitor setup. If there is any synchronized external symbology, it is added into the video. The video polarity is selected by deciding to invert all of the bits or not. This is the final step before the fully processed, 8-bit video leaves the CPLD and enters the custom analog ASIC device. Inside, it is converted back to analog with a custom 8-bit digital to analog converter (DAC) running at pixel rate. This device uses an internally generated reference voltage, and outputs a signal that is consistent with RS-170 voltage levels. Finally, the synchronizing pulses are added to the waveform to complete the RS-170 format, and a 75-ohm driver is included to allow the use of standard monitors.

6 Performance Results

Upon completing this design task, the prototype sensor was built, and many performance tests were conducted. The system was extensively verified over a wide ambient temperature range, -50 to +60 degrees Celsius (°C). The system NETD was measured using multiple techniques to obtain verifiable and repeatable results. This data is shown in Fig. 7.

Although the signal portion is measured similarly with all the methods, the noise-measurement technique varies. Some include a three-dimensional noise measurement that combines the horizontal and vertical spatial noise with the temporal noise to determine the NETD, while other measurements consider only the temporal noise component. Also, when a diffusing chopper is used, the spatial frequency of the target used is crucial to properly measuring the signal magnitude. These differences lead to small variations in the results. It is also important to note that full detector-limited performance is realized, which verifies that no unexplained degradations in

System MRTD Test Results
15 deg HFOV

Freq. (cy / mrad)	Freq. (f / f0)	Opaque Chopper MRTD (deg C)	Silicon Chopper MRTD (deg C)
0.623	1	0.200	0.109
0.312	0.5	0.039	0.053
0.156	0.25	0.012	0.013

Fig. 6. The TRP Prototype System MRTD Results

System NETD Test Results
15 deg HFOV

- Detector NETD: 0.056 C [1]

- Plastic Opaque Chopper

    Manual System NETD: 0.055 C [2]

    SBIR 'Automatic' System NETD: 0.052 C [3]

    System NETD using HIDATS2 SW & frame-grabber: 0.047 C [4]

- Silicon Diffusing Chopper

    SBIR 'Automatic' System NETD: 0.07 C [5]

Notes:
1. 11882.102 standard HIDATS2 AT on 7/10/96 -- temporal only
2. Oscilloscope method, average of (4) independent observers, 1 degree C target
3. SiTF method from -3deg to +3deg in 1deg increments
4. 2 degree C delta -- temporal only
5. SiTF method from -1deg to +1deg -- Chopper phase not optimum

Fig. 7. The TRP Prototype System NETD Results

performance exist at the system level or are caused by the electronics. Fig. 6 shows one set of minimum resolvable temperature difference (MRTD) curves. This figure demonstrates the effects of the diffusing chopper as well as the spatial noise problems of using an opaque chopper.

7 Summary

By developing this prototype sensor, the TRP made a significant impact on the potential success of RSC's uncooled FE products. Since it is only a prototype implementation intended to allow iterative development of electronic concepts, some aspects make it inappropriate for production hardware. For example, the cost, power, and size of this configuration are prohibitive. Nevertheless, the system demonstrates the feasibility of many enhanced capabilities and provides a development path to future systems with lower cost and power, smaller size, and higher performance along with several new features.

As a result of the recent merger of several companies and their respective uncooled IR technologies into RSC, many new opportunities and challenges lie ahead. Compatibility with existing products and the selection of future technical directions, which strive for commonality, are key to RSC's success. Current commercial and automotive programs promise to further advance the technologies with new finer-pitched monolithic detectors, low-cost IR optics, and further advancements in electronics in order to develop additional military and commercial products. At the same time, production is ramping up to higher volumes on existing products. These events will combine to position RSC as the leading IR systems operation in the world.

Acknowledgements

The author wishes to thank the many people at RSC who have laboured long to make this technology successful over the past two decades. Specifically, Kevin Sweetser, Barry Teague, and Ed Ramsey have had the most profound effect on this project's outstanding accomplishments. Also, a special thanks goes out to our sponsors at NVESD and DARPA for many years of support and encouragement.

Background Information

This research was funded in part under the Technology Reinvestment Project. The views and conclusions contained in this document are those of the authors and should not be interpreted as representing the official policies, either expressed or implied, of the Research and Special Programs Administration, the Advanced Research Projects Agency or the U. S. Government.

References

[1] K. Sweetser, "Infrared Imaging with Ferroelectrics", Integrated Ferroelectrics, Vol. 17, pp. 349-358, 1997.

[2] H. W. Neal, R. J. S. Kyle, "Texas Instruments Uncooled Infrared Systems", Texas Instruments Technical Journal, Vol. 11, No. 5, pp. 11-18, 1994.

[3] C. Hanson, H. Beratan, R. Owen, M. Corbin, and S. McKenney, "Uncooled Thermal Imaging at Texas Instruments", SPIE Proceedings, Infrared Detectors: State of the Art, Vol. 1735, pp. 17-26, 1992.

3.1.6 Laserscanner for Obstacle Detection

U. Lages

IBEO Automobile Sensor GmbH
Fahrenkroen 125, 22179 Hamburg, Germany
Phone: +49/40/64587-170, Fax: +49/40/64587-109
E-mail: ul@ibeo.de

Keywords: laserscanner, obstacle, tracking, object outline, mounting, nearfield

Abstract

The upcoming automotive applications for driver assistance focus on the near field environment of the vehicle, especially to the front and the side areas. For realisation of these applications, it is necessary to get a complete model of the environmental situation, consisting of all relevant objects.

In this paper the IBEO Laserscanner Technology for object detection and tracking in the nearfield of the vehicle will be presented. The presented results are based on the Laserscanner LD Automotive, which is a one plane Laserscanner with a fine resolution in distance and angel and a wide horizontal field of view. Furthermore, the development of the Laserscanner Sensor and the tracking algorithms, especially for classification are mentioned.

1 Technical Data of the Sensor

The IBEO Laserscanner has a horizontal field of view of up to 270° and up to 40 m on targets with an reflectivity of 5%. The maximum radial range is limited at 100 m. The smallest object which can be detected has an opening angle of 1°. This leads to the fact, that all relevant objects can be detected. The accuracy of distance measurement of each shot is +/- 5 cm (1 sigma) without any statistical improvement. The angle resolution is 0,25°.

The divergence of 5 mrad avoids gaps in the field of view. The scan frequency is 10 Hz, which makes it possible to track all objects with a good accuracy in lateral and longitudinal relative velocity in the near field. The sensor is eyesafe with laserclass 1.

Basic algorithms for object detection and object tracking can be integrated in the sensor on a DSP (Digital Signal Processor). For the use of the sensor with extended algorithms for object detection and object tracking (e.g. tracking of partly hidden objects; tracking of more than 20 objects) an Industrial PC as an embedded system is provided.

Fig. 1. 3 Laserscanner LD Automotive Mounted on a VW-Polo [7] (on the Left), 1 Laserscanner LD Automotive Integrated in the Bumper of an Alfa Romeo [1] (on the Right)

2 Obstacle Detection

The IBEO Laserscanner can be mounted at the front end of the vehicle in order to supervise the area in front of the vehicle, including the neighbouring lanes. In this area (field of view), all relevant objects can be detected and tracked. Each object is described by a number of variables (see Fig. 2):

- Relative Position in x and y of the "characteristic points"
- Relative Speed in x and y of the target, represented by its centre (B)
- Object Outline given by a "number of points on the outline"
- Priority for each object (object identifier related to a chosen sorting criteria – e.g. TTC)
- Object classification (under development)
- Object identifier number (1...255)

An object can be identified as an obstacle, if the collision with this object is unavoidable. Therefore, it is necessary to predict the relative moving directions in x and y, which is a typical task for a pre-crash sensor in the near field. Based on the chosen sorting criteria, the sensor provides a list of objects, relevant for the current application.

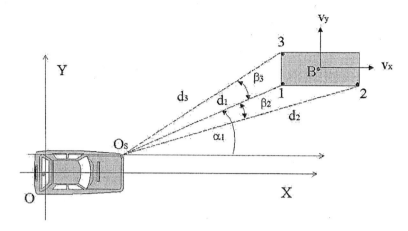

Fig. 2. Object Description of IBEO Laserscanners

Fig. 3 shows an a cyclist, avoiding the collision with a car on its lane. Therefore, the cyclist enters the lane in front of the test vehicle. On the right side of Fig. 3, the Laserscanner results of this scenario, by the use of extended algorithms for object tracking are presented. The single measurements (raw data) are represented by the single dots and the object descriptions are represented by 3 characteristic points per object, connected by two lines. It can be seen, that the IBEO Laserscanner is able to detect both objects and to distinguish between the car and the cyclist.

Fig. 3. Nearfield Driving Scenario in the City with a Car and a Bicycle [3]

3   Technology Development

The current problem of the LD Automotive is based on the fact, that this Laserscanner is a one plane sensor. If the test vehicle accelerates or decelerates, the leading objects at a distance of more than 25 m could get lost, because of the pitching angle.

Therefore, IBEO Automobile Sensor GmbH started the development of a Laserscanner with 4 planes in parallel. The first approach was made in the MOTIV project in 1997-1999. The results were so promising, that the development was extended in 2000, supported by the EU (e.g. [1]).

The new IBEO Laserscanner LD Multilayer is based on the multi-plane technology and the multi-target capability, which allows to separate at least 2 values of distance in one single shot/channel. The vertical field of view in the driving direction is 3,2 (see Fig. 4). The scan frequency can be changed between 10 Hz (e.g. for stop&go) and 40 Hz (e.g. for pre-crash).

Fig. 4.   Vertical Field of View of the Laserscanner LD Multilayer

In Table 1 the LD Automotive and the new Laserscanner LD Multilayer are preliminarily described with their characteristic technical data and the algorithms for object detection and tracking, including additional functionality like road detection.

4   Conclusion

The IBEO Laserscanner has a fine resolution in distance and angle. Furthermore, the wide horizontal field of view allows the calculation of relative movement of objects in the near field of the vehicle in longitudinal and lateral direction. This leads to a complete supervision in the field of view without any gaps. In this field of view, all relevant objects can be detected and sorted, depending on the requirements of the application. Critical objects, which will collide with the test vehicle can be classified as obstacles.

The current work is focused on the miniaturisation of the sensor and the optimisation of the algorithms for object classification, especially for pedestrian detection.

References

[1] Langheim, J.; Buchanan, A.; Lages, U.; Wahl, M.: CARSENSE - New environment sensing for advanced driver assistance systems. Proceedings of IV 2001, IEEE Intelligent Vehicles Symposium, IV 2001, Tokyo (www.carsense.com).

[2] Fuerstenberg, K.; Baraud, P.; Caporaletti, G.; Citelli, S.; Eitan, Z.; Lages, U.; Lavergne, C.: Development of a Pre–Crash sensorial system – the CHAMELEON project. Joint VDI/VW congress, Wolfsburg, 2001.

[3] Fuerstenberg, K.; Willhoeft, V.: Pedestrian Recognition in urban traffic using Laserscanners. 8th world congress on Intelligent Transport Systems, Sydney, 2001.

[4] Kirchner, A.; Lages, U.; Timm, K.: Speed Estimation with a Laser Rangefinder. Proceedings of ICARCV'96: 4th International Conference on Control, Automation, Robotics and Vision, 04. - 06.12.1996, Singapore, pp. 1875 - 1879.

[5] Lages, U.: Collision Avoidance System for fixed Obstacles. IEEE 4th International Conference on Intelligent Transport Systems, Oakland, 2001.

[6] Lages, U.; Fuerstenberg, K.; Willhoeft, V.: Der Laserscanner und seine Möglichkeiten zur Entlastung des Fahrers. VDI/SAE/JSAE Gemeinschaftstagung "Der Fahrer im 21. Jahrhundert", Berlin, 2001.

[7] Weisser, H.; Schulenberg, P.; Bergholz, R.; Lages, U.: Autonomous Driving on Vehicle Test Tracks: Overview, Motivation and Concept. IEEE International Conference on Intelligent Vehicles, Stuttgart, 1998.

3.1.7 Infrared Camera for cAR - ICAR
a EURIMUS Project for Driver Vision Enhancement

J.P. Chatard[1], X. Zhang[2], E. Mottin[3], J.L. Tissot[3]

[1]SOFRADIR
43-47 rue Camille Pelletan, 92290 Châtenay-Malabry, France

[2]UMICORE IR Glass
Z.A. du Boulais, 35690 Acigne, France

[3]CEA LETI
17 rue des Martyrs, 38054 Grenoble Cedex, France

Keywords: microbolometer, focal plane array, uncooled infrared sensor, chalcogenide, IR lens

Abstract

High potential is expected from advanced microsystems to develop low cost and high performance systems for automotive applications and a great challenge for the next few years is to improve vehicle safety during night and difficult weather condition using affordable Driver Vision Enhancement (DVE) systems. In EURIMUS frame work, a project named ICAR, is under progress to develop a specific camera for this application. This paper present first results on infrared focal plane array which will be integrated in the camera and infrared glass characteristics which will be used for the camera lens.

1 ICAR Project Presentation

ICAR (Infrared Camera for cAR) is an EURIMUS project to develop an affordable Driver Vision Enhancement (DVE) systems. The partnership is composed of CEA (F), CEDIP (F), CRF (I - Research Centre of FIAT), SOFRADIR(F - project leader), VALEO (F), UMICORE IR Glass (F) and ZEISS Optronics (D). The aim of this project is the development by the year 2003 of an all-European fully integrated low cost automotive camera for driver vision enhancement. SOFRADIR and CEA / LETI develop a new Infrared Focal Plane Array (IRFPA), UMICORE IR GLASS develops a new infrared lens material. ZEISS optronics is in charge of lens design and camera box and electronics realisations. CEDIP develops the IRFPA proximity electronics and CRF will adapt the HMI and the camera in a car taking into account the equipment specifications given by VALEO. The starting date of the project is 01/01/2001 and the duration is 30 months.

The use of infrared systems for high volume and low cost applications could be now foreseen thanks to the development of two new technologies which are the uncooled infrared focal plane array and infrared glass. We present below the IRFPA and low cost Infrared transmitting material developments after one year of project activity.

2 Uncooled Microbolometer IRFPA

2.1 Introduction

Uncooled infrared detectors are based on the thermal detection [1] of incoming IR flux instead of the quantum mechanism used in high performance cooled detectors like those made with CdHgTe or InSb. The infrared flux is detected at each pixel site by measuring an absorber temperature increase due to the amount of energy absorbed in the structure. The difficulty is then to integrate these functions in a small pixel pitch keeping a high thermal insulation between the readout integrated circuit and the absorber in close contact with the thermometer.

2.2 Microbolometer Development for DVE

The requirements of enhanced vision driver application is mainly constrained by objective cost of the system. In order to fulfill the DVE requirements we have decided to reduce the focal plane area by decreasing the number of pixels from the current 320 x 240 available devices with a pitch of 45 µm to 160 x 120 pixels with a pitch of 35 µm. To maintain a high level of performance despite the decrease of the pixel pitch, we enhance the microbolometer performance by increasing the thermal insulation and reducing the electrical 1/f noise in the amorphous silicon thermometer.

The classical embodiment of our technology exhibits a thermal resistance up to 14.10^6 K/W. An improvement of thermal resistance by a factor 3 has been obtained with a new pixel design and ultimate design rules for test pixels designed in the project framework on 35 µm pixel pitch in test structures. Concerning noise, it is well known that amorphous silicon exhibits 1/f noise and the amount of 1/f noise is linked with the volume of material used for electrical conductivity. We optimized the pixel structure to meet the Hooge law [2] requirements which described this phenomenon. As far as we have a great margin with the thermal time constant, thermal capacitance increases is still acceptable as long as the thermal time constant remains compatible with video frame rate. A rms noise reduction of 60% is obtained with this new pixel design.

Characteristics of this new technological stack are summed up in the following Table 1:

320 x 240 pitch: 35 μm	Rth 10^6 K/W	τth (ms)	NEDT (mK)
Standard technology	14	2	180
Enhanced technology	42	12	36

Table 1. Standard and Enhanced Technology Performance Comparison for 35 μm Pitch Devices

Besides, the microbolometer focal plane arrays need to be packaged under vacuum in a complex package with an infrared window. For ICAR device we switched from the standard metallic one used for 320 x 240 devices, for a ceramics one. This new package is under development and Fig. 1 shows a schematics CAD view.

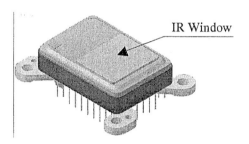

Fig. 1. Schematic View of Microbolometer Package

An advanced packaging technique is also under development, based on wafer level integration [3] for further package size and cost reduction for future devices.

3 Chalogenide Glass

3.1 Introduction

The development of low cost infrared camera involves not only the development of low cost IR focal plane, but also the development of low cost IR lens. Usually IR lens are made from germanium or zinc selenide mono crystalline ingots which are cut into blocks and then polished to obtain IR lens. This method is long and expensive and a new approach is under development for manufacturing IR lens from IR glass. This type of material could be directly molded into the final lens shape decreasing the resulting cost of manufacturing keeping a high performance level by using Fresnel lens or aspheric lens.

3.2 IR Chalcogenide Glass Development

Umicore IR Glass S.A. has developed a new chalcogenide glass composition mainly for automotive application. This glass is fabricated by using high purity germanium, antimony and selenium in a completely sealed silica ampoule. This process ensures a good reproducibility of glass quality and refractive index.

Table 2 shows some properties of this glass, called GASIR2, compared to the mostly used infrared transmitting materials for thermal imaging.

Properties	GASIR2	Ge	ZnSe
Density (g/cm^3)	4.70	5.23	5.27
Young's Modulus (GPa)	19	103	70
Poisson's ration	0.27	0.28	0.28
Thermal expansion coefficient (K^{-1})	16×10^{-6}	6×10^{-6}	8×10^{-6}
Specific heat (J/g.K)	0.34	0.31	0.34
Thermal conductivity (W/m.K)	0.25	60	16
Upper use temperature	200°C	80°C	High
Refractive index at 10.6 µm	2.6	4.0	2.4
dn/dT (K^{-1})	0.6×10^{-4}	4×10^{-4}	0.6×10^{-4}
Dispersion (8-12 µm)	101	865	58

Table 2. Physical Properties of GASIR 2 Glass, Compared with the Two Commonly Used Infrared Transmitting Materials

To obtain low-cost infrared optics, the following three additional important conditions should be fulfilled: low-cost starting materials, inexpensive process for ingot production and inexpensive process for lens fabrication. GASIR2 glass contains 20% of germanium and the other 80% of starting materials are relatively cheap. Consequently, the cost of bulky GASIR2 glass is significantly lower than that of germanium and zinc selenide.

Lenses in GASIR2 glass can of course be produced by polishing or single point diamond turning. But, the main competitive advantage of this glass is associated with the fact that even very complex lenses (asphero-diffractive for example, see Fig. 2) can be produced directly by molding, eliminating the costly polishing or single point diamond turning used for other infrared lens material. This unique infrared lens production technique by molding is particularly cost-saving for medium and high volume production, especially for complex lenses.

Fig. 2. As-moulded Asphero-diffractive Lens in GASIR2 Glass (Diameter 60 mm)

4 Conclusion

High potential is expected from advanced microsystems to develop low cost and high performance systems to improve vehicle safety during night and difficult weather condition using affordable Driver Vision Enhancement systems. In ICAR project, (under EURIMUS frame work), we are developing a specific camera for this application. The first results in focal plane and material lens developments will be integrated in the design of the infrared camera components which will be available at the end of this year and test in real conditions in the first half of 2003.

References

[1] "LETI/LIR's Amorphous Silicon Uncooled Microbolometer Development", J.L. Tissot et al, SPIE Vol. 3379, Infrared Detectors and Focal Plane Arrays V, 1998.

[2] IEEE transactions on electron devices VOL 41 n°11, 11 November 1994.

[3] "High reliability and low cost uncooled microbolometer IR focal plane array technology for commercial application", J.L. Tissot et al. (AMAA 2001), Sven Krueger, Wolfang Gessner Editors, pp209-220, ISBN 3-540-41809-1-

3.1.8 Micro-Opto-Electro-Mechanical-Systems (MOEMS) in Automotive Applications

F. Ansorge, J. Wolter, H. Hanisch, H. Reichl

Fraunhofer Institute Reliability and Microintegration
Micro-Mechatronic Center
Argelsrieder Feld 6, 82234 Wessling, Germany
Phone: +49/8153/9097-500, Fax: +49/8153/9097-511
E-mail: ansorge@mmz.izm.fhg.de

Keywords: MEMS-Packaging, Micro-Mechatronics, Micro-Mirrors

Abstract

Micro-Mechatronics & System Integration is a multidisciplinary field, which offers low cost system solutions based on the principle of homogenizing system components and consequent elimination of at least one material, component or packaging level from the system. These system approaches show, compared to the existing solutions, a higher functionality, more intelligence and better reliability performance. The number of interconnects necessary to link a motor, sensor, or an actuator to a digital bus decreases, as the smartness of the devices increases.

The paper presents problem definitions and solutions for optical devices using micro-mirror reflector technology for mechatronic systems. To reduce package volume, advanced packaging technologies as Flip Chip and CSP could be used, for increased reliability and additional mechanical functionality encapsulation processes as transfer-molding, a combination of transfer- and injection molding are recommended.

The mechatronic packaging will be discussed in detail. Especially cost, reliability performance and according "Design for Reliability" show the potential of micro-mechatronic solutions in automotive and industrial applications.

3.1.8 Optical Micro-Opto-Electro-Mechanical-Ssystems (MOEMS)

1 Introduction

The development of electromechanic assemblies has so far been carried out mainly by different manufacturers who go their individual methodology. Assembly and fine adjustment of the individual components and their integration into a final system normally takes place at the very end of the manufacturing process.

Due to more system intelligence many electronic systems comprise an additional control unit which receives information from different sensors and selects the individual electromechanic actors. These central control units need a lot of space and are fairly inert due to the long transmission paths of the electric signals. On top of that, the cabling of all the different components involved in the assembly process can be a complex and expensive business; it is definitely a major headache for mobile applications.

Sub-systems on the other hand, which mechatronically integrate both actor and sensor facilities as well as all information processing (software) in the component itself, deliver an optimised use of existing space and higher speeds of data transmission between actor/sensor functions and µ-processor technology. All this results in substantially higher performance capabilities. This would reduce transmission distances and data volumes. It would also direct the individual systems towards greater independence and autonomy. Both arguments are crucial for optical devices.

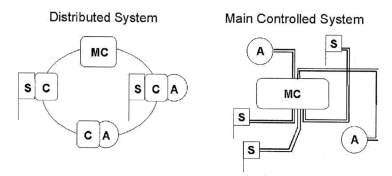

Fig. 1. Juxtaposition of a Sub-system with Bus Technology (Left) and a Conventional Control Unit (Right). (MC Main Controller; C Controller; A Actor-Mirror; S Sensor)

The overall system is characterised by the communication of the individual sub-systems both with each other and the central control unit (main controller), Fig. 1. This process can be conducted via accident-insensitive bus systems. A field bus system connecting all sub-modules with the main controller or a (thus defined) master module in the form of a ring or a string is already enough to ensure the free flow of all information.

The basic sensor data were already processed and converted in the sub-module. Only the data required by the control logic system is transmitted to the main control unit. As a consequence, you have electromechanic systems with "subsidiary intelligence" – a mechatronic system.

This multi-directional feedback mechanism results in a dominant technology, based on subtle analysis of the functions needed and optimum separation of functionality of the parts involved. Mechatronics is defined here in this sense as the application of intelligent sub-modules. A more comprehensive definition would define it along such lines as the following: Mechatronics is the synergistic integration of mechanical engineering, electronics and complex control logic into development and working procedures, resulting in mechanic and optic functionality and integrated algorithms [1], [2].

2 Basic Conditions for the Development of Mechatronics

The development of mechatronics is a vital key for the future of electronics. Ever more complex applications require the processing of ever increasing data volumes: a corresponding increase in flexibility and functionality is the only possible response developing micro-mechatronics. Synergistic co-operation between the individual departments concerned is also an indispensable condition. Electronic and mechanical simulation and designs must be compatible for a proper realisation of the thermomechanical concept (see Fig. 2). The package developed fits to an optimum place within a macroscopic system of mechanical parts, which are especially designed to work with electronics as intelligent sub-system. It is of equal importance to generate a software map of all requirements and functions, if at all possible already at the planning and development stage; this is the ideal way to shorten development times while optimising the system eventually produced. To yield this added functionality within one package the effort of specialists from various fields of technology as physics, chemistry, electrical engineering, microelectronics, software design, biology, medicine etc. are needed.

3.1.8 Optical Micro-Opto-Electro-Mechanical-Ssystems (MOEMS)

Mechatronic Design

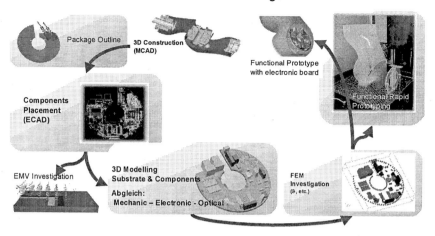

Fig. 2. Mechatronic Design Procedure

2.1 Principles of Micro-Mirror Devices

A micromirror device is a micromechanical spatial light modulator which is presently mainly used in display systems for projectors. The micromirror device is a MEMS device that uses the advantages of both IC and mechanical system. A typical system is built up on a SRAM CMOS. Texas Instruments is one of the main supplier of such DMDs (digital mirror device), other suppliers are Fraunhofer IZM Chemnitz and Fraunhofer IMS Dresden. The DMD is a reflective, passive, spatial light modulator composed of an array of rotatable aluminium mirrors currently targeted for the projection market with an resolution of up to 1280x1024 pixels. The micromirror element at the DMD is compromised of an aluminium mirror suspended over an air gap by two thin, post-supported, mechanically compliant torsion hinges that permit a mirror rotation of +/- 10°. As a digital light switch, the DMD operates under dark field projection optics to direct light from a projection source into or out of the pupil of a projection lens. The activation of the micromirror results from the landing of control voltage between mirror and grid. Due to the electrostatic force the mirror will be deformed. The profile of this deformation and the light modulated characteristics depend on the special mirror design. Generally there are two different designs, the 1-dimensional and the 2-dimensional mirror, that can be built up with a simple torsion hinge, as a pyramids element or a sag element. Light modulators with micro mechanical mirrors offer a wide range of applications, e.g. in display technology (video, data projection), in information technology (optical image and data handling, optical memory), in medical technology (laser scanning tomographie, laser surgery, endoscopy head-up display) or in manufacturing technology (laserablation and -manufacturing....) [7], [8], [9].

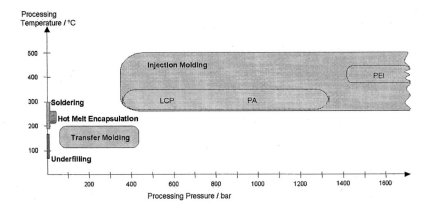

Fig. 3. Process Window of Thermoplastic and Thermoset Materials

2.2 Packaging Concepts and their Aspects on Polymer Materials for the Use in Automotive Applications

Focusing on micro-mechatronic applications there are a few additional demands to encapsulants commonly used for packaging. The encapsulants need a wide range of temperatures, a high resistance against harsh environment and the integration of moving and sensing elements without losing package functionality. Additionally for the use of micro-mirror devices clear compounds are crucial [7].

Typical materials used for microelectronics encapsulation are epoxy resins, where the chemical basis is a multifunctional epoxy oligomer based on novolac resins. These materials do have Tg's beyond 200°C and so they have the potential for short term / high temperature application. The evaluation of encapsulants for optimised long-term stability is one of the topics the micro-mechatronics centre is focusing on. Clear compounds are also already available, but due to the lack of fillerparticel there the CTE is much higher than for standard materials. This needs more sophisticated selection of package outline and arrangement of the components.

Further potential for micro-mechatronics lies in the use of advanced thermoplastic materials. In the past they have been used rather for electronics housing than for the direct encapsulation of microelectronics. The use of thermoplastic materials not only allows the integration of additional mechanical functionality (plug housing, clip on mounting, ...)

3.1.8 Optical Micro-Opto-Electro-Mechanical-Ssystems (MOEMS)

simultaneously with microelectronics encapsulation, but these materials do not cross-link and thus do have a high potential for recycling.

Research of Fraunhofer IZM-MMZ is performed in the fields of direct encapsulation of microelectronics using thermoplastic polymers, thermoplastic circuit boards a.k.a MID-devices for both, Flip Chip and SMD components. Combination of different types of materials, tailored for the application, is a key for highly reliable modules.

These applications are also capable for connecting the package geometry with mechanical functions. Only thermoplastic materials are suited for the creation of scoop-proof connectors or clip-on-connections. Their suitability for the process of direct encapsulation, is subject to a number of conditions. In order to preserve a multifunctional package, both technologies would have to be joined in the forming of two- or multi-component-packages.

Additional areas of research important for the manufacturing of miniaturized mechatronic systems are wafer level encapsulation and the handling of thin and flexible base materials. The use of thermoplastic hotmelt materials for micro-mechatronic packaging allows a decrease of processing temperature, the use of cost effective low temperature materials. [8]

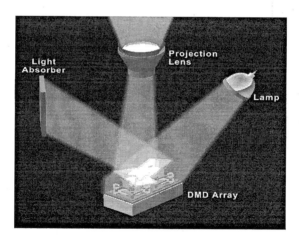

Fig. 4. DMD – Functional Sketch (source: Texas Instruments)

3 Micro-Mechatronic Applications Containing Micro-Mirror Devices

Three main areas for the application of mechatronic sub-systems in combination with Micro-Mirror technology are explained. The products of these areas have one thing in common: they consist of extensive overall systems with complex control units. Depending on the complexity of the application in question, the mechatronic design needs a defined design and interconnection technology.

The following examples illustrate the potential range of intelligent sub-systems in existing and future automotive applications. They also show the most important today's trends.

3.1 Micro-Mirrors Used for Front Light Systems

A new application for micro mirror devices will be the front light system for automobiles. In the field of the automotive-illumination (front headlight) already crucial improvements of illumination were achieved by the development of "Xenon"-lamps. In contrast to traditionally automotive-illumination, these lamps have an electronic regulation whereby the change between low beam light and high beam is no longer realised by switching between two filaments. Thus there is potential for an electronic control. A brand-new functionality is the control of the lamp reflector while driving along curves and subject to the view-direction of the driver (or by speech control). Thus an additional illumination of the roadside can be achieved which is suggestive especially in pretended dangerous situations. For this a part of the reflector mirror from the front headlights can be controlled so that the roadside which is aimed at by the driver will be additionally illuminated short-time.

Additionally functions of the reflector can be obtained by the use of micro-mirror technology. Due to the activation of single segments the driver will be able to image signs on the road by his front light systems. (e.g. arrows, road information, etc.) The functional integration in a mechatronic housing for the production of good value mass products is an essential component for the development of a new interactive security concept "headlights". Key aspects are the harsh environment at the assembly area. The temperature of the micro mirror device can be reached up to 150 C due to the under-the-hood environment. Therefore a cooling strategy for the mirror especially for the mirror control seams to be very important. Another requirement of the system is a good stability against the vibrations of the automobile. First tests already showed that the DMD is very insensitive due to the compact design.

3.1.8 Optical Micro-Opto-Electro-Mechanical-Ssystems (MOEMS)

The main attention has to be turned onto the mirror control that consists of standard ICs and resistors with a low temperature resistance. Thus it is of great importance to work out new concepts for a suitable package to avoid thermal and thermo –mechanical fatigue of the device. The target for the nearer future will be a light system with an integrated micro mirror system that guarantees a high reliability under harsh environment for budget price with the highest functionality.

3.2 Micro-Mirrors Used for Observation Systems

Another large application field for micro mirror use in automobile will be the observation technology. At this different functions are possible. Presently the existing parking aids are standard ultrasonic parking assistances. This special feature enhances the driver's perception at the rear and the front. The system uses an array of four ultrasonic sensors to help alleviate the blind spot immediately behind the rear bumper. The effective field of view is from about 10 inches off the pavement up to the top of the trunk with a maximum range of approximately five feet (source: Cadillac). The use of an optical system by micro mirrors will an efficiently new solution that might effect several advantages opposite to ultrasonic. With an micro mirror system the observation area can be specified and changed by software according to the requirements. Another advantage of a micro mirror system is the possibility to use the system also for additional observation functions. Due to the flexible scanning area of a micro mirror the system offers an enlargement of observation range than ultrasonic. Another application of micro mirrors is the pre-, rear- and side-crash control. With the use of intelligent systems the safety for the occupants and pedestrian will increase. For such applications it's of big importance to select small and flexible solutions that can be integrated easily and invisible in the car body. The needed package has to be robust and offers the possibility of intelligent connectors and switches for high comfort and high reliability.

Another operational area for micro mirrors is the blind spot detection. A great number of collisions emerge due to the blind spot. To avoid such accidents a steady observation of this area will be important. With the use of micro mirror the blind spot area can be scanned and analysed quickly.
This application requires however the highest integration of the system, a great miniaturisation and a high reliability for the assembly area.

All applications for observation purpose have to be constructed for the use in harsh environment under highest and lowest temperatures, for maximum vibrations, in an optimising package with highest system integration and universal operational area. The system has to be controlled on-the-spot as an autonomous unit to avoid long communication and minimised failures.

3.3 Micro Mirror Systems for Head Up Displays

In former times investigations regarding the use of micro mirror devices for head up displays in automotive applications are already realized. The profit of such systems is an important part of the safety control. The car manufacturer Cadillac installs an available Night Vision system -- the infrared technology allows to see up to five times farther than the low-beam headlamps can reach.

Fig. 5. Night Vision Control (Source: Cadillac)

Night Vision uses an infrared technology to "light up" the road ahead. A special camera-like sensor in the grille picks up infrared energy being radiated from objects directly ahead, processes the data in real time, and translates it into a video image. The image is then projected onto your windshield, just to the left of centre, through an Eye Cue Head-Up Display atop the instrument panel. The image sits in a natural location for the driver. The result is an enhanced view and potentially increased reaction time, which can mean greater safety when travelling at night. Night Vision fuses design and technology to give a glimpse of the future. By the use of Micro-Mirror a real time video image can be submitted I high quality onto the windshield. Due to the flexible positioning of the mirrors it is possible to adjust the image according to the driver.

The main requirements for such system are a small weight and a miniaturised system. But also a functional packaging must be taking into consideration with regard to the acceleration of the car and the high temperatures which can be reached in the cockpit. An optimised package solution should include an intelligent plug in connector for a multi functional field of application regarding the different types of cars.

3.4 Micro-Mirror Systems: General Solutions for Automotive Systems

The use of micro mirror technology in automotive applications will cause high demands on the needed system. The first step has to be the modification of the existing micro mirror application. As already mentioned the main item for the application field "automotive" will be the harsh environment. Current mirror-applications need to fulfil standard specifications of home entertainment equipment (e.g. room temperature, no vibrations or other mechanical stress etc.). Therefore new packaging solutions for the surrounding "automotive" have to be developed. According to the assembly area the main criterion will be the small size, the weight, the temperature resistance, the resistance against vibration and pollution, an easy assembly, an easy change, an easy retrofitting, a easy adaptation for several car types ... . For example the working area of the mirror system is of great relevancy. If the use of the developed system will be near by the motor housing the achievable temperature can reach up to 150° C – a value at which most of the electronic components will crash down. Therefore a cooling system has to be taken into consideration. Also the kind of package must be well selected to avoid any fatigue of the material. Another example could be a blind spot system integrated in the doors. At this application the temperature has not the critical effect but the mechanical steadiness. The forces effected by closing the doors can reach the highest values at the car. Therefore this assembly area has again other emphases for the packaged system. The mirror has to be possibly integrated in a damped surrounding which has to be considered regarding the chosen package (e.g. package material, package process...).

This short summary shows the importance of individual solution according to the field of application. Presently there are a lot of interesting applications in automotive sector. In near future the micro mirror technology will increase in all industrial areas but the automotive applications will lead the way.

4 Outlook

Future fields of application for mechatronics will include optical systems for near field observation mechanisms. Table 1 reflects the requirement profiles of the individual application fields. The trend towards mechatronics persists in all these fields and is even gaining momentum. According to research studies, a large part of what is today still a vision of the future we also gave you a glimpse of that in our paper will be converted into reality by micro- machine technology already in the period between 2002 and 2007: the integration of sensorics, actoric functions and controllers will make it possible.

Requirement / Use	Reflector Types	Head-Up Displays	Observation Systems	Summary
Share of Sensorics	++	++	++	++
Miniaturisation	+	+++	++	+
Modularity	+++	☐	+	++
Cost effective	+++	++	+++	+++
Market potential	↗	↗	↗	↗

Table 1. Mechatronics requirements and trends (+++ crucial, ++ very important, + important, ☐ optional, ↗ increasing).

Acknowledgement

Parts of this work are funded by the "Bayerisches Kompetenznetzwerk fuer Mechatronik" part of the "High Tech Offensive Zukunft Bayern".

References

[1] Control and Configuration Aspects of Mechatronics, CCAM Proceedings; Verlag ISLE, Ilmenau, 1997.

[2] F. Ansorge, K.-F. Becker; Micro-Mechatronics – Applications, Presentation at Microelectronics Workshop at Institute of Industrial Science, University of Tokyo, Oct. 1998, Tokyo, Japan.

[3] F. Ansorge, K.-F. Becker, G. Azdasht, R. Ehrlich, R. Aschenbrenner, H.Reichl: Recent Progress in Encapsulation Technology and Reliability Investigation of Mechatronic, CSP and BGA Packages using Flip Chip Interconnection Technology, Proc. APCON 97, Sunnyvale, Ca., USA.

[4] Fraunhofer Magazin 4.1998, Fraunhofer Gesellschaft, Muenchen.

[5] F. Ansorge; Packaging Roadmap – Advanced Packaging, Future Demands, BGA's; Advanced Packaging Tutorial, SMT-ASIC 98, Nuremberg, Germany.

[6] Toepper, M. Schaldach, S. Fehlberg, C. Karduck, C. Meinherz, K. Heinricht, V. Bader, L. Hoster, P. Coskina, A. Kloeser, O. Ehrmann, H. Reichl: Chip Size Package – Michael A. Mignardi; "From ICs to DMDs", TI Technical Journal, July-September 1998, pp. 56-63.

[7] Long-Sun Huang; "MEMS Packaging for Micro Mirror Switches", Abstract, University of California.

[8] Jeff Faris, Thomas Kocian; "DMD Package – Evolution and strategy"; TI Technical Journal, July-September 1998, pp-87-94.

[9] Fraunhofer Institut Mikroelektronische Schaltungen und Systeme; "Lichtmodulatoren mit mikromechanischen Spiegelarrays".

3.2 Powertrain Applications

3.2.1 Overview: Microsystems as a Key Eelement for Advanced Powertrain Systems

A. Noble[1], M. Voigt[2], S. Krueger[2]

[1]Ricardo Consulting Engineers Ltd, Bridge Works, Shoreham-by-Sea
West Sussex BN43 5FG, United Kingdom
Phone: +44/1273/794378, Fax: +44/1273/794563
E-mail: adnoble@ricardo.com

[2]VDI/VDE-IT
Rheinstrasse 10B, 14513 Teltow, Germany
Phone: +49/3328/435-277, Fax: +49/3328/435-256
E-mail: voigt@vdivde-it.de
Phone: +49/3328/435-221, Fax: +49/3328/435-256
E-mail: krueger@vdivde-it.de

Keywords: application, powertrain, pressure sensor

Abstract

This paper will give an overview on current sensors and actuators being relevant for the powertrain system. It also introduces challenges for future microsystems, therefore addressing objectives which help to increase flexibility, economy and environmental compatibility of engines.

1 Introduction of System Applications

Demands continue to escalate on automotive powertrains for cost-effective improvement of fuel economy and emissions whilst meeting increasing customer expectations for refinement and drivability. In order to meet these demands engines are becoming increasingly complex with a growing electronics content, with variable valve train, intake system, port deactivation, variable geometry turbocharging and so on. There are also a growing number of operating modes making the control more complex, for example lean stratified, lean homogeneous, homogeneous rich, regeneration, warm-up modes in gasoline direct injection engines. Figure 1 gives an overview on state of the art microsystems being integrated in the powertrain system.

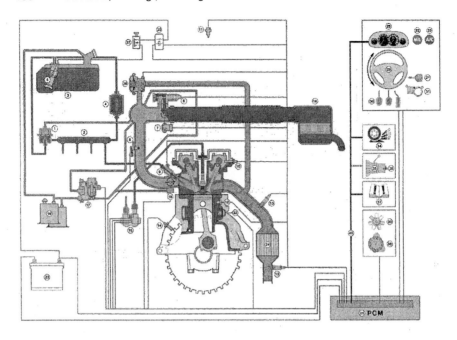

1	Fuel-pressure controller	21	Ignition switch
2	Fuel distribution strip	22	Thermostat
3	Tank	23	Battery
4	Fuel filter	24	Three-way catalytic converter
5	Fuel pump	25	Engine control unit
6	Idle control valve	26	Electric exhaust loop valve
7	Throttle valve sensor	27	Fuel safety valve
8	MAP and temperature sensor	28	Instruments
9	Injection valve	29	Servo-pumps pressure switch
10	Camshaft position sensor	30	Clutch pedal switch
11	Environmental temp. sensor	31	Immobiliser
12	Coolant temperature sensor	32	Malfunction display
13	Heated lambda sensor	33	Air condition control
14	Crankshaft position sensor	34	Anti slipping control
15	Ignition coil	35	Transmission
16	Active carbon vessel	36	Speed sensor
17	Active carbon flush valve	37	Diagnostic system
18	Air filter	38	Charging current control
19	Tumble flaps	39	Fan control
20	Fuel-pump relay	40	Connector to multiplex wiring

Fig. 1. Powertrain System (by courtesy of Visteon)

2 Powertrain

As shown in Fig. 1 basic clusters of the powertrain systems are the fuel, the intake, the combustion, the aftertreatment system and the transmission. Each of the systems incorporates a set of different sensors being used for control functions. There are major differences depending on the fuel used (e.g. gasoline or diesel) and the required characteristic of the engine leading to specific design solutions and additional subsystems.

The objective behind the powertrain system is, starting with sufficient fuel, to produce the exact amount of power (speed and torque) needed for the instantaneous specific driving situation to move or accelerate the automobile. There are conflicting needs, for instance sacrificing fuel consumption to have a cleaner combustion, or increasing the comfort by using an automatic transmission with ever present pressures to decrease the price. Microsystems are especially a solution for control tasks, meaning to get the relevant information in the ever changing processes and reacting by changing boundary conditions or ratios of the contribution substances.

The idle, transition and full-throttle systems all contribute to matching performance to programmed curves, describing the different situations of the engine. Further secondary functions may include; ignition control, transmission-shift control, fuel consumption displays and diagnosis capabilities. Emissions control is divided into two basic categories; engine design measures (A/F ratio – lambda, mixture formation – quality of A/F mixture, uniform distribution in all cylinders, exhaust gas recirculation, valve timing, compression ratio, combustion chamber design, ignition system, crankcase ventilation) and exhaust gas treatment (thermal afterburning, catalytic afterburning, catalytic converters). The standard closed loop three way catalytic converter requires very precise lambda control.

- Lambda Sensor (oxygen-concentration sensor)

Can be rendered useless by lead that may be present in the fuel or exhaust gas.

The fuel-metering system emplays the exhaust gas residual-oxygen content as measured by the lambda sensor to very precisely regulate the air-fuel mixture for combustion to the valve lambda. At its easiest the sensor is a solid-state electrolyte (Zirconium-dioxide). Electrically heated sensors are specially well-suited for the measurements in the lean range. Wide-range lambda sensors of multilayer ceramic design can also be used in diesel engines.

- Combustion Sensors

Combustion pressure sensors are not routinely used in automotive engine management but interest is growing for their use in the future. Controlling the pressure, temperature and the spread of the combustion substances is the key issue for engine optimisation, to deliver a smooth, clean combustion and high power. Cylinder pressure, fiber optics, ion-current-sensing sensor elements are direct ways to get these needed information. The problem for these systems is the harsh environment in the combustion chamber where temperatures up to 1.000°C and pressures up to 15 bar are reached. For detailed information please refer to the AMAA 2000 book, chapter on engine control systems.

- Knock Sensors

Knock sensors measure the structure-borne noise at the engine block (approx. 10 g at a typical vibration frequency of 5...20 kHz) for an electronic control of the ignition point. Even if the ignition angle as a function of engine speed, load and temperature is controlled it is necessary to ensure that even in the most knock-sensitive case with regard to engine tolerances, engine ageing, environmental conditions and fuel quality, no cylinder reaches or exceeds the knock limit. Engine compression, fuel consumption and torque improves. Knock sensors, being acceleration sensors are inexpensive systems and could become obsolete if combustion sensors are adopted in the future.

- Position Sensors (displacement and angle)

The throttle valve, the accelerator pedal, the fuel level and the travel of the clutch servo unit are examples for systems being observed by sensor systems. For these applications position sensors might be used. So far contact wipers (potentiometers being most economical) or contactless Short-circuiting ring sensors are the most economical solution. Half-differential sensors, hall sensors, magneto-resitive sensors and optical systems have the potential to replace these systems thanks to better reliability and accuracy.

- RPM and Velocity Sensors

Examples for systems using these sensors are the crankshaft and camshaft speeds, wheel speed and the speed of the diesel injection pump as well as the speed of the turbo fans. Here inductive, Hall effect and optical micromechanical sensor systems might used.

- Pressure Sensors

Intake manifold pressure (1...5 bar), modulator pressure for automatic transmissions (35 bar), fuel tank pressure for on board diagnostics (0,5 bar), combustion chamber pressure for misfire and knock detection (100 bar, dynamic), diesel pumping element pressure (1000 bar, dynamic) and electronic diesel injection, common rail pressure (1500...1800 bar) for diesel engines, common rail pressure (100 bar) for spark ignition engines are areas where pressure sensors are used in powertrain systems. There is a wide range of systems and specific solutions available. Please refer to the chapter pressure sensing in this book for detailed information.

- Flow Meters

For knowing the exact amount of air being used for combustion flow meters are implemented in the induction pipe. While hot wire solutions were used a long time, microsystems sensors working with thin membranes on silicon are state of the art. It is possible that accurate cylinder pressure sensors that can estimate trapped air in each cylinder during the intake stroke will render mass air flow sensors obsolete.

- Torque Sensors

Torque being the most important value for the car and its movement is so far calculated by different values, e.g. from the composition of the mixture fed to the combustion chamber. For this purpose data on the inducted air mass, the pressure in the air induction pipe or the angular position of the throttle valve are determined. A direct measurement that is inexpensive and reliable has so far not find its way into the automobile. First promising solutions are suggested – one being introduced later on in this almanac.

- Temperature Sensors

Temperature sensors monitor the outer temperature, temperatures of the air and the fuel used, the engines operating temperature. They are also implemented in other sensor systems for computational reasons, meaning that for instance a pressure sensor does not behave linear over its full range and temperature dependent formulas in the electronic control unit assure an accurate signal of the whole sensor system. The electrical resistance of metals, semiconductors and conductive ceramics changes with temperature, allowing some of these materials to be used as thermo sensors.

- NOx Sensors

The major proportion of the exhaust gas is composed of three components – nitrogen, carbon dioxide and water vapour. The minor components are carbon monoxide, nitrogen oxides, hydrocarbons and particulates. For engine control reasons especially for lean burn and gasoline direct injection engines NOx sensors are more often used but adding costs to the system.

It remains important to notice, that due to millions of units of the named sensors being produced, the highly automated production process and low quantities of materials used, microsystems solutions are opening the way for advanced control systems while the powertrain system stays competitive on price.

3 Outlook

The engine will remain the most important component of a vehicle. It is for all OEM's crucial to at least fulfil governmental regulations as well as customers demands. Nevertheless the powertrain is regarded by OEMs as a key subsystem that defines the character or 'brand DNA' of the vehicle.

Improvements that can be made by microsystems come under the following categories:

- Throttle actuation, where better smoothness and precision and very high reliability are necessary in the future torque-based gasoline engine management systems.
- Fuel actuation, where extremely accurate control of timing, quantity and rate is necessary in the latest generation common-rail diesel engines. Spray and air motion estimation and control are also critical in diesel and direct injection gasoline engines.
- More precise measurement of engine position.
- Measurement of cylinder pressure with more robust, accurate and durable sensors.
- Measurement of knock has a long history in gasoline engines. Now it is finding its way into diesel engines with, for example, the Delphi system using an accelerometer for monitoring start of combustion and closed loop control of pilot injection.
- Measurement of torque at various points in the powertrain, for example crankshaft, transmission input and output shafts and drive shafts.
- Electromagnetic or electrohydraulic valve actuation where control of valve opening and closing angles and lift, possibly with position feedback and very rapid actuation.
- Exhaust gas composition, historically monitored in gasoline engines by lambda (oxygen) sensors which are now also finding their way into diesel engines. Monitoring of new exhaust gas constituents such as oxides of nitrogen, carbon monoxide and hydrocarbons may well be a requirement in more tightly controlled powertrains of the future.
- For hybrid powertrains, new requirements are emerging for precise current and voltage measurement and control.
- Oil condition sensing or diagnostics of materials and media in general will become a key element for expanding service cycles while remaining reliable.

- High temperature electronics as well as sensors are needed to be fitted into or closed to the interesting points e.g. the combustion chamber, the exhaust pipe or the transmission oil.

It remains open, which Microsystems solutions will find their way into the powertrain. Some interesting applications will be introduced in the following articles.

References

[1] S. Krueger, W. Gessner, "Advanced Microsystems for Automotive Applications 2002", Engine Control session

[2] Bosch, Automotive Handbook, 5th Edition, Robert Bosch GmbH, 2000

3.2.2 A Piezoelectric Driven Microinjector for DI-Gasoline Applications

C. Anzinger, U. Schmid, G. Kroetz, M. Klein

EADS Deutschland GmbH, Department SC/IRT/LG-MX
81663 Munich, Germany
Phone: +49/89/607-29969, Fax: +49/89/607-24001
E-mail: claus.anzinger@eads.net

Keywords: actuator, direct-injection, gasoline, injector, piezoelectric sensor, stroke amplifying unit

Abstract

This paper introduces a new microinjector designed for optimising gasoline direct-injecting engines. The basic idea for the construction of this microinjector was to develop a new kind of valve with an extremely high lifting speed. Generally, the latter can be realised by integrating a piezoelectric stack actuator in the injector in combination with small masses for the accelerated parts. Due to the material properties of piezoceramics, this solution requires an actuator which is comparatively long in size. But it is desirable to keep the actuator as short as possible in order to reduce the production costs and to additionally increase the lifting speed of the valve. To adapt the specifications of the actuator to the requirements of the valve, this present microinjector operates with a new, miniaturised stroke amplifying unit. The miniaturisation allows the integration of the unit into the injector tip without extending the outside diameter of 7 mm. Simulations have been performed to demonstrate the high dynamic performance of the injector as well as to optimise the stiffness of the mechanical parts that are directly involved in the opening/closing procedure. Injection quantity measurements have also been performed with the first prototypes, using a hydraulic test bench.

1 Introduction

The standards for economical automotive fuel consumption as well as reduced exhaust emissions are rising and hence, the demands for new concepts in the field of the injection technology. In gasoline engines, it is possible to use the advantages of direct injection into the combustion chamber, similar to the diesel direct injection [1]. But significant differences in these two combustion processes require varying injection systems.

3.2.2 A Piezoelectric Driven Microinjector for DI-Gasoline Applications

In this paper we present a novel injector for gasoline direct injecting engines. Because of its miniaturised setup it is called microinjector. A prototype of the latter is shown in Fig. 1.

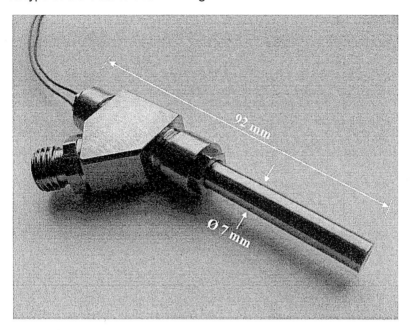

Fig. 1. Photograph of the Microinjector

In comparison to the standard systems used until now, the microinjector allows a faster opening and closing of its valve needle. Only this quality makes it possible to take the maximum advantage of the direct injection by injecting highly precise and a repeatable amount of fuel per cycle [2]. In addition, the application of a considerably higher fuel pressure supports an optimal spray atomisation [3].

Conventional solenoid injectors cannot meet these requirements. Therefore, we aimed at a new concept for an injector, to contribute to the suggested advancements. The microinjector is based on a piezoelectric stack actuator. This kind of actuator is able to expand and contract itself in an extremely short time. One of the outstanding features of the microinjector is its novel, miniaturised stroke amplifying unit. This sophisticated unit intensifies the movements of the piezoelectric driven actuator and, hence, it compensates for the inherent disadvantage of this kind of actuator – namely the small stroke. The result is a new injector, working up to 150 bar, with an opening and closing time of less than 0.1 ms. The outer diameter at the injector tip is 7 mm, and the overall length is about 92 mm [4].

2 Basic Considerations for Designing the Microinjector

2.1 Features and Advantages

Piezoelectric stack actuators consist of lead-zirconium-titanate-ceramics (PZT), which expand in electric fields. This effect proceeds in an extremely short time interval and creates high mechanical forces. The actuator in the microinjector can apply up to 300 N in less than 0.1 milliseconds [5]. The maximum strain is 0,1% of the actuator length [5]. This maximum value is reduced by the stiffness of a supplementary prestressing mechanism, which is necessary to avoid tensile stress in the piezoceramics. Tensile stress occurs due to the high accelerations in the actuator and could damage the brittle piezoceramics. Therefore, the effective stroke of the prestressed actuator in the microinjector amounts to only 80% of the maximum value. The piezoelectric actuator and its installation in the microinjector are shown in Fig. 2.

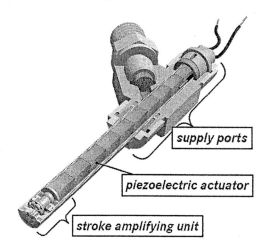

Fig. 2. Schematic View on the Arrangement of the Microinjector

Calculations and tests have shown, that the lifting height of the valve needle needs to be about 140 µm in order to obtain a homogenous, tapered fuel spray at the required delivery rate. In contrast to other piezoelectric injection valve concepts, based on very long piezo stacks, this microinjector uses a shorter actuator for the same needle lift. The difference in length is compensated by a sophisticated stroke amplifying unit with a multiplication factor of about three. Consequently, in this configuration, the microinjector needs an actuator of only 70 mm length, instead of more than 200 mm. The actuator employed is a standard multilayer low voltage stack with a maximum/effective stroke of 70 µm / 55

3.2.2 A Piezoelectric Driven Microinjector for DI-Gasoline Applications

µm at 150 V, which equals an electric field strength of 2 kV/mm. Its cross-sectional area is approximately 12 mm² [5], and, as the leads are integrated in the housing, the diameter is constrained to 6 mm.

The following points sum up the basic advantages of a smaller actuator, and thereby the most outstanding benefits of the microinjector configuration:

- The size of the piezoelectric actuator is decisive for the production costs. Especially for injector mass production, a smaller actuator is favoured, as the price of a piezoelectric stack is approximately proportional to its length, assuming a constant diameter.
- The length of the actuator determines its resonant frequency [5] and, as a result, the switching time of the injector. Consequently, a faster injector needs a shorter actuator.
- The electrical properties of piezoceramics are similar to those of a capacitor. A change in the strain of the piezo stack is equivalent to the charging or discharging of a capacitor, thus requiring high currents for a fast expansion of the actuator. Short-term powers can climb up to 50 A and more. The microinjector with the shorter piezo stack, however, allows the use of a controller with a lower output power.
- The recharging of the actuator is associated with power dissipation [6]. Again, a larger actuator produces higher losses. The dissipation is converted into heat, which has a negative impact on the stroke. Because of the low thermal conductivity of the ceramic material, the worst case scenario is the damage of the actuator due to overheat.
- When expanding the actuator, high forces occur as a result of the extreme acceleration. To avoid buckling of the piezo stack, the ratio of length to diameter is preferably designed not to be too high. The resulting greater diameter of a longer actuator causes a higher and disadvantageous capacitance.

But especially the first two benefits can be reduced by a poor design of the stroke amplifying unit. To avoid any influence reducing the resonant characteristics of both, the actuator and the whole system, the force-transmitting parts of the amplifier have to be as stiff as possible, in combination with small masses for the accelerated parts.

2.2 The Stroke Amplifying Unit

The stroke amplifying unit of the microinjector is located between the piezoelectric stack actuator and the outwardly opening valve needle, as shown in Fig. 2. It operates mechanically on the basis of lever transmission and multiplies the stroke of the actuator at the expense of the available force.

The most important component of this unit is the amplification diaphragm, basically a plane circular metal membrane of a diameter of about 5 mm and a distinct structure. This structure consists of eight levers, arranged in a circle and fixed to each other at the circumference. At its outer diameter, the diaphragm is fixed in the axial direction at the housing of the amplifying unit. The valve needle is fixed at the tips of the lever arms. In between these fixing points lies the circular point of application of the piezoactuator, whereby this positioning depends on the desired transmission ratio. Fig. 3 gives a detailed view into the setup and the operating principle of the stroke amplifying unit.

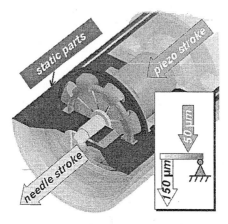

Fig. 3. The Operating Principle of the Stroke Amplifying Unit. The different intensity of amplification is illustrated with corresponding greyscales - black means static and white represents the maximum amount of amplification.

One advantage of this configuration is the high stiffness of the transmission mechanism due to the eight parallel levers. Another benefit is the symmetry of the setup and the symmetric power flow through the whole injector. This is designed so as not to deflect the microinjector asymmetrically and thereby not to influence the spray-cone.

The design of the membrane is defined by two main considerations. On the one hand, it is important to keep the compliant connections between the single levers weak enough to avoid stress concentrations and a loss of amplification. On the other hand, the levers have to be as stiff as possible in order not to substantially affect the dynamic behaviour of the injector. The optimal combination of these requirements has been determined by using the well established FEM software tool ANSYS 5.6. These simulations show, that it is possible to combine the required characteristics in a plain, structured membrane without getting too high stresses or too low stiffness. This result is important for the manufacture of the amplification membrane as a low-cost part. Various production methods are possible, especially for its structuring. Our experiments with prototypes of this component have revealed problems with thermal methods like laser cutting because of distortion. But we believe that techniques like electro erosion or micromilling are suitable solutions, (compare Fig. 4). The fabrication as a cheap stamped part would only be appropriate for mass production.

Even the bearing and the housing of the diaphragm allow a small outside diameter of the whole stroke amplifying unit. It does not exceed the corresponding diameter of the actuator housing. The components of the bearing, the housing and the valve have been manufactured by means of micromachining as well as by precision conventional machining. Therefore, it was possible to integrate the amplifying unit into the microinjector without extending the diameter defined by the piezoelectric actuator.

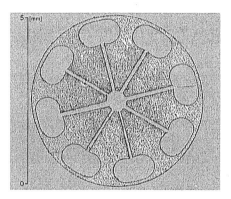

Fig. 4. Photograph of a Laser-cut Amplification Diaphragm

2.3 Extra Design Considerations

Because of the elasticity of the materials, fuel pressure fluctuations in the injector lead to different strains and compressions both in the actuator and the housing parts. Such variations in length have to be taken into consideration and are balanced through an appropriate layout. The same applies to thermal effects, since the coefficient of thermal expansion of piezoceramics is significantly smaller than that of the steel housing.

2.4 Dynamic Simulations

The dynamic simulation of the mechanical arrangement of the microinjector without fuel made clear, that the needle lifting time is characterised by the oscillatory characteristics of the actuator, as shown in Fig. 5. The stroke amplifying unit does not excite any relevant additional oscillations, and its transmission of the actuator movement to the valve needle is approximately proportional. With the damping being substantially higher in a fuel filled microinjector, the undesired oscillations die down even faster.

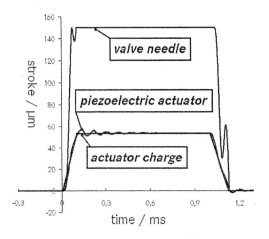

Fig. 5. Dynamic Simulation of the Mechanical Arrangement of the Microinjector, Disregarding any Influences from the Fuel

If it is intended to further reduce lifting times as well as oscillations, additional methods, like using an optimised non-linear charging and discharging curve, could be applied to the microinjector.

2.5 Measurements with the First Prototypes

The first experiments with the prototypes of the microinjector were performed statically in order to check the function of the stroke amplifying unit (see Fig. 6). The actuator can expand by up to 55 µm, depending on the applied voltage. At the same time, the valve needle is displaced by 135 µm. The stroke amplification factor is about three. The offset in the beginning of the charging is caused by a free travel of the actuator before the needle moves. Therefore, a closing force between the valve needle and the valve seat is induced, which is needed for a leak-proof closing of the injector. Of course this is associated with a minor loss of the effective needle stroke, but this fact is unavoidable, even with an unamplified arrangement.

For the hydraulic evaluation of injection systems, volume measurements have to be performed at an injection test bench. Therefore, a complete injection system is set up with all motor boundary conditions. An overview

3.2.2 A Piezoelectric Driven Microinjector for DI-Gasoline Applications

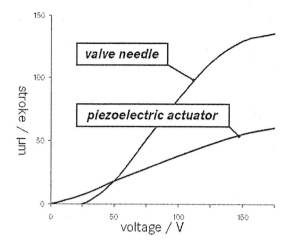

Fig. 6. Strokes of the Piezoelectric Actuator and the Valve Needle, as a Function of the Applied Voltage to the Actuator

of this test bench is shown in Fig. 7. It can be subdivided into the following components:

- Injection system
- Electric motor (drive for the high-pressure pump)
- Sensors for number of revolutions, rail pressure, needle stroke and temperature
- Control unit for the injection system
- Test bench control system
- Injection amount indicator (EMI)
- Measurement data recording (IMAD)

The main component of the injection test bench is the injection amount indicator (EMI). It determines the injected fuel quantity of each stroke temporally dissolved. The amount of fuel, which is injected by the nozzle, displaces a piston inside the EMI. As the displacement of the piston is proportional to the injected fuel quantity, the latter can be determined by measuring the position of the piston. This sensing is done by an water cooled, inductive displacement sensor. On the rear side of the piston, a nitrogen pad exists to simulate the compression pressure arising in the engine. After each injection, the measuring chamber is emptied, using a drain valve. The measured signals are transferred to the measurement data recording (IMAD). It controls the test bench and is connected with the control unit of the injection system. Because of the coupling of the test bench with the injection system, it is possible to perform injection quantity measurements with characteristic maps.

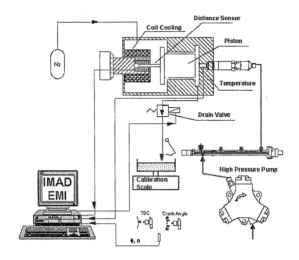

Fig. 7. Schematic Overview of the Injection Test Bench

A subsequent differentiation of the measured signal yields the dynamic injection rate, as shown in Fig. 8.

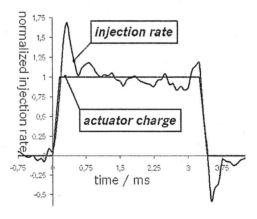

Fig. 8. Injection Rate of the Microinjector (Fuel Pressure 130 bar, Test Oil Shell 1434). The injection rate is related to the static flow rate.

Fig. 8 shows the direct response of the injection rate to the charging of the actuator. In comparison to this, a typical injection rate diagram of a conventional injector is marked by a delay between the control signal and the beginning of the injection [7]. Time intervals of 100 µs are here the lower limit.

3 Conclusion

A highly sophisticated concept for a microinjector is presented in this paper. It is realised technically by the means of precision mechanics. The advanced characteristics of this injector are the small dimensions, the high dynamic response and the piezoelectric actuator technology combined with a miniaturised stroke amplifying unit. The small size, especially the diameter of the injector tip, facilitates the application of the microinjector to automotive multi-valve cylinder heads. It is even possible to imagine the installation of two microinjectors per cylinder in order to realise two different spray-cone characteristics for mean- and full-load running. Further research needs to focus on experiments concerning the spray atomisation qualities of the microinjector and also reliability tests. The results of this experience will show if and where the concept needs to be optimised. We are, however, confident that the microinjector contributes positively to a further development of the direct injection system for gasoline engines.

Acknowledgements

The authors wish to thank D. Mehlfeld (Dept F2/EA, DaimlerChrysler AG), T. Flämig-Vetter (Dept EP/VRA, DaimlerChrysler AG) and S. Krell (Temic) for valuable discussions and their support.

References

[1] C. Stan (Hrsg.): Direkteinspritzsysteme, Berlin, Heidelberg, 1999.

[2] H. Eichlseder, E. Baumann, B. Hoss, P. Mueller, S. Neugebauer:, Herausforderungen auf dem Weg zur EU IV-Emissionserfuellung bei Otto-Direkteinspritz-Motoren, 21. Wiener Motorensymposium, 2000.

[3] S. Kraemer: Untersuchungen zur Gemischbildung, Entflammung und Verbrennung beim Ottomotor mit Benzin-Direkteinspritzung, VDI Fortschrittsbericht, 1998.

[4] C. Anzinger, U. Schmid, G. Kroetz, S. Krell: Microinjector For DI Gasoline Engines Based On MST-Actuator, MST News, 2001.

[5] Piezoelectrical and Electrostrictive Stack Actuators, Product Catalogue, Piezomechanik GmbH, 2001.

[6] J. Heinrich: Modellierung und Simulation piezokeramischer Aktoren als Grundlage zur Entwicklung alternativer Steuer- und Regelprinzipien, Isle, Ilmenau, 1999.

[7] K.-H. Hoffmann, K. Hummel, T. Maderstein, A. Peters: Das Common-Rail Einspritzsystem. MTZ Motorentechnische Zeitschrift 58, 1997.

3.2.3 A Robust Hot Film Anemometer for Injection Quantity Measurements

U. Schmid, G. Kroetz

EADS Deutschland GmbH, Department SC/IRT/LG-MX
81663 Munich, Germany
Phone: +49/89/607-20657, Fax: +49/89/607-24001
E-mail: ulrich.schmid@eads.net, URL: www.eads.net

Keywords: common rail, high pressure, hot film anemometer, injection rate, injection system, on board diagnosis

Abstract

In this study, a micromachined hot film anemometer, placed on a pressure stable LTCC-substrate, is presented to measure the different injection quantities, needed in modern direct injection (DI) system for optimum performance. A bi-layer of Titanium (Ti) and Platinum (Pt) as advanced metallisation system for the thin film resistors is used due to a lower resistivity and a higher temperature coefficient of resistance (TCR) compared to Molybdenum (Mo), which was taken for the first tests. Especially the increase in TCR is recommended for thermal mass flow sensors, as the resulting signal height is linearly related to this material parameter. Therefore, the new technology steps, to fabricate the sensor elements, are given in the paper. The latest injection rate measurements, performed for the first time up to 90 MPa with a nozzle integrated mass flow sensor, yield a strong increase in sensor amplitude at higher rail pressures. After the closing of the nozzle, pressure waves, caused by the strong deceleration of the fuel column, are clearly detectable, demonstrating the quality of the sensor signals.

1 Introduction

Silicon micromachining is a key technology to fabricate fast responding, low-cost and low-power sensors operating according to the thermo-resistive measurement principle. The basic idea is to detect via the temperature coefficient of resistance (TCR) of the sensing element the change in heat flow rate to or from the ambient atmosphere. Especially on the field of mass flow sensors for gaseous media, the so called hot film or hot wire anemometry principal has proven its excellent performance and is well established as a reliable technique for measuring even turbulent

3.2.3 A Robust Hot Film Anemometer for Injection Quantity Measurements

and/or unsteady flows [1], [2]. For liquid applications, however, only very few studies are reported in literature, focusing mainly on water as test medium at low fluid pressures [3]. The latter is due to the high mechanical forces, which would act on these fine cylindrical or membrane structures being in the µm or sub-µm range respectively. In contrast to all these investigations, the requirements for thermal flow sensors integrated „on board" into the injection system of an automobile to resolve as precisely as possible the injection rate is completely different. First, injection quantities of gasoline or diesel fuel have to be measured during the full life cycle of the vehicle. Second, the high fuel pressures, ranging in Common-Rail (CR) injection systems during full load operation up to 135 MPa for diesel engines [4], require a stable substrate for the electrically heated thin film resistor. In a previous study [5], we have for the first time successfully demonstrated, that a micromachined mass flow sensor can be integrated into the body of a CR injection nozzle for monitoring the injection rate. In this paper, however, the reasons for the use of the advanced metallisation system of Titanium (Ti) and platinum (Pt) for the flow sensitive layer are given, including the necessary changes in the fabrication process. Finally, the latest injection rate measurements with the improved sensor metallisation up to 90 MPa are presented and discussed.

2   Device Fabrication and Electrical Characterization of the Ti/Pt Metallisation

The basic technology steps, reported in [5], to fabricate the sensor chip are for the new metallisation system as far as possible kept to save development costs. Therefore, the LTC (Low Temperature Cofired) ceramic is first mechanically polishing to reduce the surface roughness by more than one order of magnitude and to clean the contact area to the electrical feedthroughs. Instead of Molybdenum (Mo), which was chosen for the first tests because of its good adhesion to the glass-ceramic substrate, a bi-layer of e-beam evaporated Titanium ($\approx$10 nm) and Platinum ($\approx$100 nm) is presently used for the flow sensitive thin film element. Ti substantially increases the adhesion of the noble Pt thin film on non-metallic surfaces like silicon dioxide (SiO_2) or ceramics, if a post deposition anneal of about 600°C or higher should be avoided. To get acceptable etch rates of Pt in aqua regia of about 15 nm/min, the etchant must be heated to about 65°C. On the other hand, in the temperature range below 70°C, a too strong networking of the photo resist (AZ 4562) is avoided and hence, it can be stripped in acetone. Finally, the sensor chips, as shown in Fig. 1, are cut out of the glass-ceramic plate with an in-house available ultrasonic milling process. The electrical properties of thin films, as the resistivity at room temperature or the temperature coefficient of resistance (TCR), strongly differ from crystalline („bulk") values due to the size effects or grain boundaries [6]. Therefore, rectangular test samples are fabricated on the glass-ceramic substrate, according to the above described technology, with a fixed length of L=5mm and variable width w, ranging with nominal values from 20 µm to 80 µm in 10 µm steps. By plotting the total resistance of each resistor R_{tot} as a function of the geometry factor L/w, the resistivity ρ can be determined according to [7]

$$R_{tot} = \rho \cdot \frac{L}{w \cdot d}$$ Eq. 1

by a least square fit procedure, as the thickness d of the thin film is controlled during the evaporation process. Subsequently, the total step height of the metallisation system is verified with an alpha step 200 profilometer from TENCOR instruments. A clear reduction in resistance and, as an equal film thickness is chosen, in resistivity for the Ti/Pt ($\rho_{Ti/Pt}$ = 1.7·10^{-7} Ω·m) by a factor of almost 5 is demonstrated, compared to the Mo metallisation (ρ_{Mo} = 8.2·10^{-7} Ωm), depicted in Fig. 2.

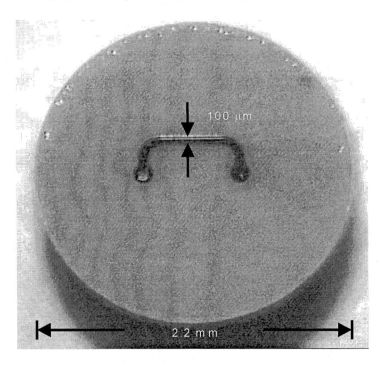

Fig. 1. Top View on a Ti/Pt Thin Film Sensor Having a Width of 100 µm on the Glass Ceramic Chip (2.2 mm in Diameter) With Two Integrated Electrical Feedthroughs

3.2.3 A Robust Hot Film Anemometer for Injection Quantity Measurements

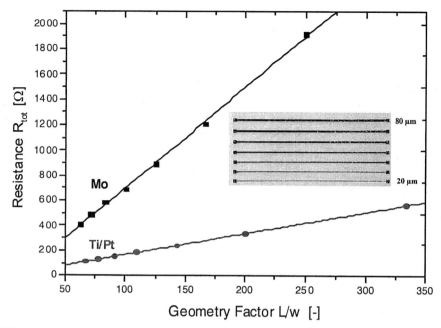

Fig. 2. A plot of resistance R_{tot} vs. geometry factor L/w, measured at different test structures on the LTCC substrate, depicted in the small picture, exhibits ohmic behavior as well as the resistivity at a 100 nm thin film. Compared to standard literature with $\rho_{Mo,lit} = 5.4 \cdot 10^{-8}$ Ω·m and $\rho_{Pt,lit.} = 1.07 \cdot 10^{-7}$ Ω·m [7], a strong increase of this material parameter is measured.

Therefore, higher sensor temperatures T_s can be achieved with the same sensor design at a given voltage supply U_s, as, according to [7],

$$T_s \propto P = \frac{U_s^2}{R_s} \qquad \text{Eq. 2}$$

where P and R_s denote the electrical heating power and the temperature dependant sensor resistance respectively. This may be necessary for a future integration in the injection system of an automobile, as the vehicle system voltage is limited to presently 12 V. On the other hand, T_s should be kept below 473 K in oil atmosphere to minimize drifts of sensor parameters during operation [8]. More important for the present investigations is the increase in TCR, as for a given decrease in sensor temperature, caused by the convective heat transfer to the cooler fluid for a fixed fuel velocity field close to the nozzle body, the signal height ΔR_s can be directly increased via

$$R_s - R_a = \Delta R_s = R_a \cdot \alpha \cdot \Delta T_s = R_a \cdot \alpha \cdot (T_s - T_a) \qquad \text{Eq. 3}$$

where Ta, Ra and α denote the ambient temperature, the sensor resistance at Ta and the TCR of the sensor metallisation respectively. Static current-voltage (I-V) measurements, performed in a vacuum chamber to minimize any oxidation of the test devices, reveal between 298 K and 473 K a $\alpha_{Ti/Pt} = 1.85 \cdot 10^{-3}$ K^{-1} in contrast of $\alpha_{Mo} = 9.3 \cdot 10^{-4}$ K^{-1}, as shown in Fig. 3.

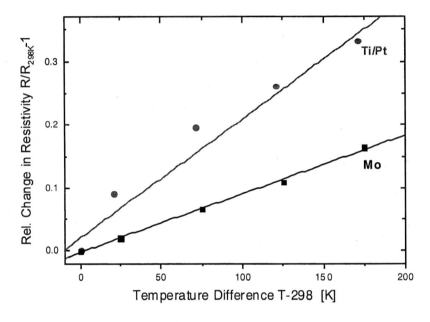

Fig. 3. Relative change in resistivity as a function of temperature difference $\Delta T =$ T-298 K of a Ti ($\approx$ 10 nm)/Pt ($\approx$ 100 nm) test structure on a LTCC substrate. Due to the linear relationship between both quantities, the temperature coefficient of resistance α can be determined, according Eq. 3.

By this new metallisation for the hot film anemometer, the sensor signal are simply doubled leading with carefully chosen electronic filters to an improved signal-to-noise ratio. Again, the values, reported for crystalline samples in standard literature [7], deviate with $\alpha_{Mo,lit.} = 4.7 \cdot 10^{-3}$ K^{-1} as well as $\alpha_{Pt, lit.} = 3.9 \cdot 10^{-3}$ K^{-1} clearly from the data given above.

In comparison, electrical material parameters of the Ti/Pt bi-layer with regard to hot film anemometry are experimentally determined, enabling a proper interpretation of the sensor characteristics.

3 Injection Rate Measurements up to 90 MPa

In this chapter, the first injection rate measurements up to 90 MPa, done with the nozzle integrated Ti/Pt hot film anemometer and performed at a high pressure test bench, described in detail elsewhere [5], are presented. Due to the effective noise reduction in the electronic readout, used for the first injection rate measurements up to 60 MPa [5], only the fixed resistors in the Wheatstone bridge are adjusted, as shown in Fig. 4.

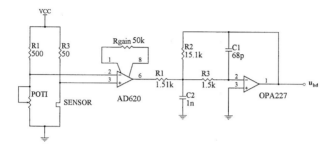

Fig. 4. Circuit diagram showing the mass flow sensor in the Wheatstone bridge and the subsequent electronic readout. The corner frequency (-3dB) of the latter is designed to 100 KHz. V_{cc} is 10V and the voltage drop across the sensor is measured to 3.70V at the operating point.

With a supply voltage for the Wheatstone bridge of V_{cc} = 10 V, a voltage drop across the thin film resistor of R_{so} = 29.3 Ω at the working point is measured to 3.70 V, what is equivalent to an electrical heating power of P = 466 mW. With the temperature dependant resistivity of the Pt film, the overheat temperature ΔT_s can be estimated with the cold resistance of the sensor R_o = 23 Ω to about 150 K, according to Eq. 3. Therefore, any influences on the sensor signals, arising from a too low temperature difference during operation, as e.g. a high sensitivity to ambient temperature fluctuations, can be minimised [9]. According to previous measurements with Molybdenum hot film anemometers, a change in u_{bd} occurs after the needle has been lifted due to the cooling of the thermal mass flow sensor by the flowing fuel. After the injection nozzle has been closed, pressure waves, originating from the fast deceleration of the fuel column at a rail pressure of 90 MPa, are clearly detectable, as depicted in Fig. 5.

Note that these information are only gained with such a nozzle integrated mass flow sensor. These oscillations in the nozzle need about 20 ms to be completely damped and the sensor reaches its previous working point after the injection pulse again. The high rise time of about 1 ms compared

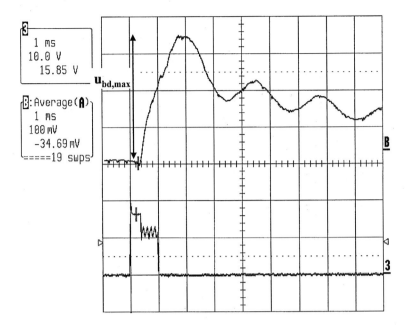

Fig. 5. Amplified (Gain G = 2) and Filtered Sensor Signal at t_{open} = 1ms with an Amplitude of About 360 mV at 90 MPa

to our previous study [5] until the maximum diagonal bridge voltage $u_{bd,max}$ is reached may be due to a non-optimised integration of the present sensor chip in respect to the nozzle body. Small changes in level height at the transition of nozzle body to the glass-ceramic chip can lead to areas of reverse flow at the position of the hot film anemometer at high fuel velocities and hence, to a less effective cooling. The latter is supported by the higher slope of the sensor characteristics, when the injection pulse starts. Detailed investigations with the EMI (Injection Amount Indicator)/IMAD system, integrated in the hydraulic test bench [5], have to be performed in the near future to verify the sensor characteristics. It is also worth mentioning, that the time delay, until the thermal flow sensor responses via the fuel mass flow to the drive pulse of the magnetic valve, is with about 300 µs in good comparison to former injection rate measurements [5]. In the pressure range between 40 MPa and 90 MPa respectively, the maximum diagonal bridge voltage $u_{bd,max}$, defined in Fig. 11, depends linearly on the applied rail pressure, as shown in Fig. 6.

3.2.3 A Robust Hot Film Anemometer for Injection Quantity Measurements

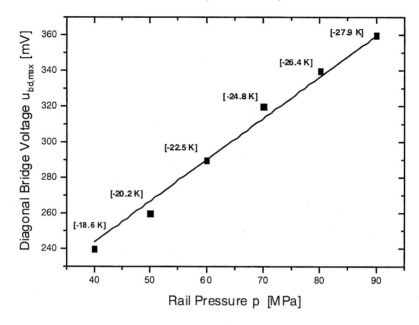

Fig. 6. Increasing $u_{bd,max}$ Due to a Higher Heat Transfer Coefficient as a Function of Rail Pressure p

The amplitude increases clearly with higher injection pressures, as higher fuel velocities occur, which lead via the higher convective heat transfer to the fuel to a higher change in resistance of the hot film anemometer. Due to all these improvements, presented so far in this study, a high sensitivity of the sensor amplitude to variations in rail pressure of about 2.32 mV/MPa result, whereas the latter is gained by a linear least square fit procedure from Fig. 6.

According to [10], $u_{bd,max}$ can be directly related to the decrease in sensor temperature during fluid flow for metallic films, having a positive temperature coefficient of resistance. With $\alpha_{Ti/Pt} = 1.85 \cdot 10^{-3}$ K^{-1}, the calculated data, ranging between 18.6 K for 40 MPa and 27.9 K at 90 MPa respectively, are additionally inserted in Fig. 6. These calculations demonstrate, that electrical material parameters, as the resistivity and the TCR, have to be precisely measured, if the signals of hot film anemometers should be properly analysed.

4 Conclusion

In this paper, a high-pressure stable hot film anemometer is presented, which allows to measure on board of an automobile, equipped with a modern CR injection system, the different fuel quantities, needed for optimum performance. Despite the good characteristics of Mo thin film sensors on LTCC-substrates in monitoring the injection rate, the performance of the device is clearly improved by using a Ti/Pt bi-layer for the flow sensitive metallic resistor. This advanced solution leads to a decrease in resistivity by a factor of 5 and simultaneously increases the temperature coefficient of resistance by a factor of 2. With well-designed filters, eliminating the electronic noise in the electronic readout, the sensor signals can be simply doubled, leading to an improved signal-to-noise ratio. Therefore, the most important material parameters for thermal mass flow applications are experimentally determined, enabling a proper interpretation of the sensor signals. Injection rate measurements, performed with the micromachined hot film anemometer in the nozzle at rail pressures between 40 MPa and 90 MPa, exhibit an increase in sensor amplitude at higher fuel pressures, leading to a rail pressure sensitivity of 2.32 mV/MPa. In accordance with former investigations, pressure waves, due to the strong deceleration of the fuel column after the closing procedure of the nozzle, are clearly measurable. Because of these precise information about the flow conditions during and after the injection pulse, we believe, that robust hot film anemometers have big potentials to optimise the combustion process of CR injection systems by a closed-loop control of the injection quantity.

Acknowledgements

The authors wish to thank Dr. G. Renner (Department EP/MDH at the Daimler Chrysler AG) and Mr. Poschner (FT1/MP) for valuable discussions and their support.

References

[1] F. Durst and A. Al-Salaymeh, "A novel single-wire sensor for wide range flow velocity measurements," 10th Intern. Conf. Sensor2001, Nuremberg, Vol. 2, pp. 41-46, May 2001.

[2] J.J. van Baar, R. W. Wiegerink, G.J.M. Krijnen, T.S.J. Lammerink and M. Elwenspoek, "Sensitive thermal flow sensor based on a micro-machined two dimensional resistor array," Proc. of the 11th Intern. Conf. On Solid State Sensors and Actuators (Transducers '01), pp. 1436-1439, 2001.

3.2.3 A Robust Hot Film Anemometer for Injection Quantity Measurements 183

[3] Bedoe, H. Fannasch and R. Mueller, "A silicon flow sensor for gases and liquids using AC measurements," Sensors and Actuators A, Vol. 85, pp. 124-132, 2000.

[4] K. H. Hoffmann, K. Hummel, T. Maderstein and A. Peters, "Das Common-Rail Einspritzsystem- Ein neues Kapitel der Dieseleinspritztechnik," MTZ Vol. 58, pp. 572-582, 1997.

[5] U. Schmid, G. Kroetz and D. Schmitt-Landsiedel, "A Mass Flow Sensor For Fuel Injection Quantity Measurements in DI-Systems," 5th Conference on Advanced Microsystems for Automotive Applications (AMAA), Springer Press ISBN: 3-540-41809-1, pp. 139-155, Berlin 2001.

[6] T.J. Coutts, "Electrical conduction in thin metal films," Elsevier Scientific Publishing Company, ISBN: 0-444-41184-4, 1974.

[7] H. Kuchling, "Taschenbuch der Physik," Fachbuchverlag Leipzig, ISBN: 3-343-00759-5, 1986.

[8] H. Eckelmann, "Hot wire and hot film measurements in oil," DISA Info., No. 13, pp. 16-22, 1972.

[9] H.H. Bruun, "Hot wire anemometry-principles and signal analysis," Oxford University Press, ISBN: 0-19-856342-6, 1995.

[10] P. Freymuth, "Frequency response and electronic testing for constant-temperature hot wire anemometers," J. Phys. E.: Sci. Instr.,Vol. 10, pp. 705-710, 1977.

3.2.4 Non-Contact Magnetostrictive Torque Sensor - Opportunities and Realisation

C. Wallin[1], L. Gustavsson[2]

ABB Automation Technology Products AB
Control & Force Measurement, Dept. FM/GM
721 59 Vasteras, SWEDEN

[1]Phone: +46/21/342498
E-mail: christer.s.wallin@se.abb.com

[2]Phone: +46/21/343095
E-mail: lars.a.gustavsson@se.abb.com

Keywords: magnetostrictive sensor, non-contact, powertrain, torque

Abstract

The Torductor®-S is a robust and durable non-contact magnetostrictive torque sensor well suited for powertrain applications in a vehicle. The design of the sensor makes it very rugged and tolerant to the demanding requirements of a vehicle and the flexibility of the sensor makes it possible to integrate with existing components in the driveline. The applications of a torque sensor in the powertrain are numerous both for engine and driveline control as well as improved diagnostics capabilities. Access to a reliable torque signal opens up for a completely new approach to powertrain control improving driveability and performance of the vehicle throughout its lifetime.

1   Introduction

During the past few years the automotive industry has realised a dramatic increase in the use of electronic systems. This development has been driven by governmental legislation for lower exhaust emission levels, reduced fuel consumption, increased performance and increased reliability. The introduction of electronic control of the engine and transmission has also led to an increased use of sensors for temperature, pressure and mass flow as an input to the electronic control system. In combination with more elaborate engine models and more refined control, the number and performance of sensors has continued to grow.

3.2.4 Non-Contact Magnetostrictive Torque Sensor

However, although sensors are used to measure a number of important parameters, there is no sensor up to now that measures the actual output of the engine measured as power or torque. Instead, control schemes for engine and transmission rely on engine models that are able to use the measurement of indirect parameters to calculate an estimated torque. Due to its nature, the quality of this calculated torque is limited and it is also dependent on variations in the engine due to fuel quality, individual variations, and wear of components. This puts a limit on the performance of the control algorithms and consequently on the performance of the engine.

With a torque sensor in the powertrain it is possible to obtain a true closed-loop control of the actual power output from the engine, independent of any uncertainties in an engine model. Furthermore, utilisation of the torque signal reaches beyond the use of overall torque. By characterising the transient properties of the torque signal the sensor can be used to monitor and control dynamic properties of both the engine and gearbox.

The reason why traditional torque sensors based on strain gauges have not been adopted by the automotive industry is mainly due to their intricate and sensitive design. They also depend on separate means to transfer the signal from the rotating shaft to the surroundings. A solution to both the demands on robustness and true non-contact measurement can be achieved by the use of magnetostrictive torque sensors. Since these sensors are based on magnetic fields they are not dependent on a direct contact with the rotating shaft. The sensing element is the actual shaft that is prepared to give a desired anisotropy. This class of sensor constitutes a very robust and durable overall solution and are regarded as a very promising technology within the automotive industry.

2 Design of the Torque Sensor Torductor®-S

The measuring principle of a magnetostrictive torque sensor is the magnetoelastic property of the load-carrying shaft. This causes a change in the magnetic permeability when the shaft is subjected to mechanical stress from an applied torque. Although based on the same physical property, a number of different principles to utilize this property have been presented over the years. Characterising features of magnetostrictive torque sensors are the orientation of the measuring coils and the preparation of the shaft surface. An overview and classification of different design concepts has been made by Fleming [1]. Several of these sensors have been presented for the measurement of engine torque e.g.[2], [3].

2.1 Operating principle

A schematic outline of the sensor is shown in Fig. 1 with the rotating sensor shaft surrounded by primary and secondary windings in the static housing. The windings are separated into three different parts corresponding to the three different zones on the shaft. The shaft is excited with a rotationally symmetric magnetic field by the primary coil.

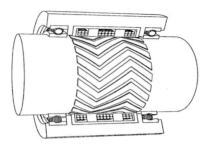

Fig. 1. Principle Drawing of a Three-zone Magnetostrictive Torque Sensor

A characteristic feature of the sensor is the thin copper pattern on the surface of the shaft. The pattern is fundamental to the function of the sensor by aligning the magnetic field with the direction of this pattern. At the three different measuring zones this pattern is arranged at ±45 degrees, which correspond to the direction of the principle stresses for a cylindrical shaft under torsional load.

When the sensor is subjected to torque the distribution of magnetic flux along the shaft is shifted due to the magnetoelastic effect. This is detected as a difference in induced voltage between the secondary coils corresponding to the applied torsional load on the shaft.

2.2 Sensor performance

Due to the low electrical impedance and lack of mechanical contact with the measuring shaft, the sensor is able to operate reliably in hostile environments such as gear boxes and engine blocks. The design with three zones of copper strips compensates for temperature effects in terms of both overall temperature [4] and temperature gradients [5]. The sensor output is a high-level modulated carrier signal, which effectively suppresses both electric and magnetic interference. A typical characteristic is a signal to noise ratio of 60 dB, even at high frequencies.

The sensor is not a permanently magnetic device and does not attract metal particles that can be found in e.g. transmission fluid. Furthermore, it does not require any special handling or care during installation, which is very important in order to avoid an increased production cost of the vehicle.

3 Installation in the Powertrain

Since the sensor measures a property that is transmitted through the powertrain it can not be installed as a bolt-on device, but needs to be integrated into actual load carrying components. For optimal result the sensor needs to be considered at the design stage of a new transmission. An important consequence of this close integration is that the sensor is located at the centre of the torque path directly measuring the primary property, without relying on indirect side-effects. This makes the measurement very reliable and repeatable.

There are no sensitive parts on the sensor shaft or the housing and the strong output signal from the sensor eliminates the need for electronic parts close to the sensor area. This makes it possible to integrate the sensor in close vicinity to heat sources in the power train, such as the engine block or clutch pack. In extreme cases, the sensor has been proven to operate at temperatures up to 250°C. Furthermore, the flexible design of the sensor makes it possible to integrate into existing components in the powertrain. The small radial space of the sensor results in a very compact overall solution.

3.1 Location of the sensor

The integration of the sensor in the powertrain has three natural locations, as indicated in Fig. 2 for a powertrain with a manual gearbox. These are between the crankshaft and flywheel, on the input shaft, and also the output shaft of the gearbox. Equivalent locations can be found for an automatic, where the flywheel is replaced with a converter. The most

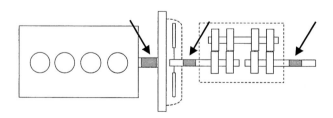

Fig. 2. Suitable Points of Installation in the Powertrain with a Manual Gearbox

suitable location for an application depends on the design of the particular powertrain and of the desired properties of the sensor. If the primary use of the torque signal is for engine control and monitoring the best location is as close as possible to the engine. In the same way, location in the gearbox is best suited for control of gearshifts. Since the different locations are subjected to different load ranges, the sensor has to be designed accordingly.

3.2 Integration to Powertrain Components

The sensor measures on the outer surface of a solid or hollow shaft and inevitably requires access to some axial length of exposed shaft. In most cases this space can be provided without compromising the overall dimensions, due to the very limited radial space requirement and the flexible design of the sensor. As an example of this, the sensor has been fitted inside the release bearing guide on the input shaft of a manual gearbox. On an automatic gearbox it has been fitted inside the stator shaft of the converter.

For a compact and robust integration, the shaft portion of the sensor can be integrated directly into the existing load carrying shaft. With the Torductor®-S, a standard shaft of case hardened or tempered steel can be used directly for the sensor function. The shaft is designed for its normal mechanical function with a middle part prepared with the characterising copper pattern. By using existing components, the integration can be carried out with a minimum of added weight.

4 Engine Applications

The application of a feedback torque signal for the engine control system has implications and potential for several subsystems in the engine. Direct torque measurement can be used to improve timing control, control of turbo charge pressure, electronic throttle, idling control, misfire detection, cylinder balancing, transient response and many more. In short, the use of a torque signal has implication on most aspects of engine control and changes completely the fundamentals of engine control. To fully meet the potential of direct torque control, new control schemes are required that take full advantage of the possibilities with a torque signal.

The benefits of using the torque sensor does not only include a refined engine control. The increased redundancy of the system reduces the accuracy requirements of other sensors and also improves the limp-home capabilities of the vehicle. Also, since the sensor gives a true feedback control, the system can adapt to the actual characteristics of the engine. In the long run, this may lead to cost savings on engine components due to reduced requirements on mechanical tolerances.

The most natural position of the sensor for engine applications is integration in the engine. For most purposes however, an installation in the driveline serves equally well.

4.1   Engine Control

Most engine control systems today can be categorised as torque based controllers, which means that the engine torque is a primary parameter of the control. Despite this, the actual engine torque is not directly available, but is calculated from a mathematical model of the engine based on input from other sensors in the engine. The accuracy of this estimated torque may be as good as 5% under ideal circumstances, but is usually much lower (may be as low as 35%) in practice due to various reasons.

The benefits of replacing this estimated value with a direct measurement are obvious. First of all, the work associated with adapting an engine model to a specific engine is simplified. Secondly, inaccuracies in the engine model due to uncontrollable parameters are eliminated. The engine maintains its characteristics throughout its lifetime independent of individual variation and wear. Another aspect is that the calibration costs of the engine in the production line can be reduced.

The introduction of direct torque measurement fits very well into the framework of existing control schemes, since they are already based on torque as a primary parameter. One approach is to simply replace the calculated torque with measured torque in a feedback control loop. An alternative way is to use an adaptive control scheme in which the engine map or engine model is continuously updated based on the measured torque. Both these approaches give the benefits of an improved control under all circumstances of the engine. They also give the possibility to compensate for wear of engine components and adaptation to different fuel quality.

In addition to an improved steady state control, direct torque measurement gives the opportunity to control and improve the transient response of the engine in a manner that has not been possible before. The response time to the driver request can be decreased without the risk of instability or overshoot, thereby giving a "sporty" feel of the vehicle. Other types of control that can be realised relates to the smoothness of the engine, such as improved idling control. It has been shown how the torque signal can be used for cylinder balancing [6] that usually requires the access to individual cylinder pressure sensors.

4.2 Engine Diagnostics

Since the torque signal shows the actual output from the engine it contains information about any unexpected behaviour of the engine. Analysis of the characteristics of the torque signal gives an insight into the condition of the engine that is not possible to reach with the existing set of sensors, since they primarily show the input to the engine. By maintaining an engine model an estimated torque can be calculated that shows the nominal performance of the engine. Deviations between this torque and the measured torque indicates that there is a fault in the engine. Analysis of the character of the deviation even gives the possibility to indicate the source of error. Continuously updating a record of this deviation gives the opportunity to monitor the state of the engine and to detect wear. A slow change in deviation over time indicates the wear of the engine, whereas a sudden change is a signal of a more acute error. Early warning of any malfunction of the engine may save large repair costs in the long run and gives the possibility to condition based maintenance.

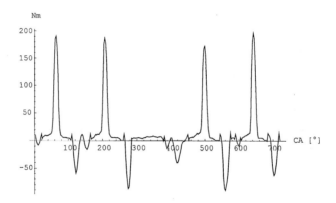

Fig. 3. Misfire Detection Using a Torque Sensor

Every single combustion leaves a typical trace in the torque signal and an analysis of the character of the torque signal reveals minute details that are related to the performance of the engine. It has been shown in [7] how the torque signal can be used for a robust misfire detection at all loads and speeds at both steady state and transient conditions even during very rough road conditions. Fig. 3 shows the torque signal for a complete engine cycle and how clearly the effect of a failing cylinder is seen in the resulting torque impulse. Other diagnostic applications that are related to uneven combustion of continuous or intermittent character can be identified equally well.

5 Driveline Applications

The torque sensor finds a number of applications in the driveline and can improve several aspects of the driveline performance in terms of closer control, faster response, and adaptation to actual driving conditions. It affects areas such as driveability, comfort levels and fuel economy due to a more optimised use of the driveline and better management of the whole powertrain. Direct torque measurement in the driveline opens up for a completely new way of controlling the driveline at a level that is not possible to achieve with current control systems. With access to a reliable torque signal the control system can perform and continuously optimise its function based on actual conditions in the driveline. Examples of specific functions in the driveline that can be improved are clutch control, gearshift strategies, tip in/out control, power management for hybrids and overload protection.

5.1 Overview of Driving Modes

The torque signal contains a wealth of information in addition to the overall torque level. By analysing the characteristics of the torque signal, detailed information of the instantaneous state of the driveline can be revealed before it is seen in any other sensors or noticed by the driver. Application of the torque signal can be used both for a refined control of the transmission and monitoring of the condition of engine components. The changing characteristics of the torque signal and how they reflect different driving conditions is best shown with an example. Fig. 4 shows the resulting torque in the driveline during a power start event for a vehicle from rest with open throttle.

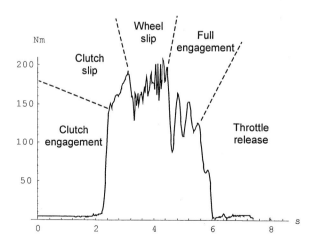

Fig. 4. Torque Characteristics During Power Take-off

The graph shows five distinct phases during the event that are clearly identified in the torque signal with a very specific signature in each single phase. The sequence begins with the rapid increase of torque when the clutch starts to engage. This is followed by a controlled clutch slip during acceleration and steadily increasing torque. A sudden drop in torque appears when the wheels start to slip, which results in a characteristic "rough" appearance of the torque signal. When the wheels regain their grip the whole transmission is fully engaged and the torque transient excites a damped oscillation. Finally, the throttle is released and the torque drops to zero.

The above discussion is an example of a driving situation where the powertrain shifts from different characteristic driving modes. With an appropriate application of the torque signal to the power train, the behaviour of the vehicle in all these modes can be improved and also the transition between the different modes.

5.2 Gearshift Control

A property of the vehicle that is crucial to the feel and impression of the vehicle is management and control of gearshifts. The application of direct torque measurement in this area provides a significant improvement over the systems that are in use today. This is especially true for AMT:s (Automated Manual Transmission) where the torque interruption and clutch engagement are very critical for the perceived smoothness to the driver. Fig. 4 shows the resulting torque from a sequence of gearshifts with open loop control and slow engagement of the clutch.

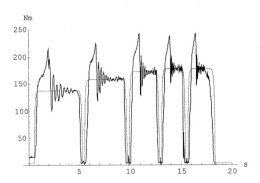

Fig. 4. Automatic Gearshift of a Manual Transmission Using Open-loop Control

The graph clearly shows that although the clutch engagement is slow it results in a large torque transient that excites a damped oscillation in the driveline. This is perceived by the driver as a "poor" gearshift and the comfort level is far from that of most automatics. With the feedback signal from a torque sensor, the clutch can be actively controlled to give a gearshift closer to the ideal response, indicated with the dashed line in the graph. Furthermore, it is possible to achieve a better synchronisation that decreases the period of torque interruption during a gearshift. For a twin clutch gearbox it has been shown by Wheals et al. ([8] and [9]) how near "invisible" gear shifts can be achieved using feedback control with a torque sensor.

5.3 Gearshift Management

An important factor for good management and control of gearshifts is the quality of the torque value from the engine control system. This is the case for AMT:s as well as for standard automatic transmissions. The estimated torque can be seen as the common interface between engine and gearbox and the quality of this value is crucial for the operation and perceived quality of the powertrain as a whole. Gearshifting on an automatic involves a well tuned synchronisation with the engine control and a lot of work during development concerns the development of gearshift strategies that gives the right "feel" to the vehicle.

The torque value that is delivered from the engine control system is calculated from an engine model and depends on a number of uncontrollable parameters such as individual variation of components, wear, and the quality of the fuel. These all contribute negatively on the accuracy and reliability of the calculated torque. The application of a torque signal that shows the actual output from the engine can be directly translated/transferred to better quality and reliability of gearshifts and gearshift management. Access to a signal from a sensor that measures the actual torque in the driveline further simplifies the work associated with adapting a gearbox to a specific engine.

5.4 Other Applications

Beside the applications that have been mentioned above, the torque sensor can be used in several other areas in the driveline. These include improved differential control and active torque distribution control for AWD vehicles. Closer control of the driveline and better adaptation to changing driving conditions leads directly to an improved off-road performance of the vehicle. The torque sensor may be further used for an active overload protection of the driveline. In the long run this can save both weight and cost by reducing the need for over dimensioning of driveline components.

For vehicles that shift between different operating modes, such as GDI engines, hybrids or ISG power trains, the torque sensor can be an invaluable key in order to improve the comfort level during the transition phase. An important issue for hybrids and ISG powertrains is to have full control over the distribution between the different power sources and to avoid the risk of torque oscillations between the individual sources. A torque sensor in the power train is in an excellent position to monitor this situation and may even be used as a tool for active damping of oscillations in these power trains.

6 Conclusions

A torque sensor concept has been developed that is able to operate reliably under the demanding requirements of a vehicle. The sensor has the robustness and durability that is required for powertrain applications combined with a flexibility that gives a compact integration. The sensor has been proven in a number of installations both for engine control and monitoring as well as driveline applications. These installations has proven the value of the sensor and shown some of the opportunities with a well incorporated use of the torque signal. The applications of the sensor ranges from diagnostics of single combustions in the engine to management of gearshifts in the driveline. The benefits from this are improved comfort levels, reduced emission levels and reduced fuel consumption.

The full potential of the sensor can not be judged only from the impact on separate functions in the powertrain, but must be regarded from a comprehensive application to the complete powertrain. A strategic application of torque measurement in the vehicle is the key that opens up for an integrated management of the whole powertrain.

References

[1] W.J. Fleming, "Magnetostrictive Torque Sensors - Comparison of Branch, Cross and solenoidal Designs", SAE Technical Paper 900264.

[2] Y. Nonomura, J. Sugiyama, Ksukada, & M. Takeuchi, "Measurements of Engine Torque with Intra Bearing Torque Sensor", SAE Technical Paper 870472

[3] R. Klauber et al., "Miniature Magnetostrictive Misfire Sensor", SAE Technical Paper 920236

[4] J.R. Sobel, D. Uggla, H. Ling, "Magnetoelastic Noncontacting Torque Transducer", PCT Patent Application 9500289, March 1995

[5] J.R. Sobel, "Magnetoelastic Torque Transducer", US Patent 4,873,874, assigned to Asea Brown Boveri, issued Oct 1989

[6] J. Jeremiasson, C. Wallin "Balancing of Individual cylinders in a V8 Diesel Engine Based on Crankshaft Torque Measurement", SAE Technical Paper 981063

[7] J.R. Sobel et al "Instantaneous Crankshaft Torque Measusrement in Cars", SAE Technical Paper 960040

[8] J.C. Wheals, L.M. Sykes, "Integrated Powertrain: Control Strategy", Ricardo International Conference, 21/06/2000

[9] J.C. Wheals et al., "Integrated Powertrain Control Applied to AMT Designs for Improved Shift Quality, Efficiency and other Characteristics Using Novel Hardware and Sensors", VDI Berichte nr. 1610, 2001

3.2.5 Benefits of a Cylinder Pressure Based Engine Management System on a Vehicle

P. Moulin[1], A. Akoachere[1], A. Truscott[1], A. Noble[1], R. Mueller[2], M. Hart[2], G. Kroetz[3], C. Cavalloni[4], M. Gnielka[4], J. von Berg[4]

[1]Ricardo Consulting Engineers Ltd, Bridge Works, Shoreham-by-Sea
West Sussex BN43 5FG, United Kingdom

[2]DaimlerChrysler AG, HPC T721
70546 Stuttgart, Germany

[3]EADS Deutschland GmbH
P.O. Box 80 04 65, 81663 Munich, Germany

[4]Kistler Instrumente AG
Eulachstrasse 22, 8408 Winterthur, Switzerland

Keywords: combustion, model based control, piezoresistive sensors, pressure sensing, silicon on insulator

Abstract

Global demands in fuel economy and lower emissions are having an increasing impact on the development of automotive Engine Management Systems (EMS). Cylinder Pressure based EMS (CPEMS) technology has attracted much attention in recent years because of the potential benefits in engine performance and reliability. AENEAS, a collaborative project between Ricardo, DaimlerChrysler, EADS and Kistler with part funding from the European Commission and the Swiss Government, has demonstrated the benefits of this technology on a vehicle. The key in realising such benefits lies in the combination of inexpensive robust sensors based on Silicon on Insulator chip technology, and model-based control and diagnostics algorithms.

This paper describes the background to this project. It provides details of the sensing technology and describes the development and implementation of control and diagnostics algorithms on a Mercedes Benz E320 vehicle. Results are included to demonstrate the benefits of this new technology.

3.2.5 Benefits of a Cylinder Pressure Based Engine Management System

1 Introduction

The constant drive for better fuel economy and lower emissions from automotive vehicles of today has led to increasingly stringent requirements in the development of Engine Management Systems. Such technology is being pushed to the limits with the need for improved control and monitoring of individual combustion events. The application of cylinder based control and diagnostics techniques have therefore attracted much attention in recent years. In particular, the use of cylinder pressure as a feedback signal offers potential improvements in fuel consumption, emissions, On-Board Diagnosis (OBD) as well as a smoother running engine. This technology is becoming more practical for automotive applications in the near future as on-board microprocessing power increases and new robust pressure sensing materials become more affordable.

AENEAS was a collaborative project between Ricardo, DaimlerChrysler, EADS and Kistler with financial support from the European Commission and the Swiss Government. Its aim was to demonstrate some of the benefits of CPEMS technology on a Spark Ignition (SI) engine [1]. This paper describes the background to this project and presents some of the results obtained.

2 Sensing Technology

2.1 Material Characterisation

For measuring the cylinder pressure in combustion engines a piezoresistive pressure sensor has been developed. Due to the high temperatures in the combustion chamber [2], [3] the piezoresistive sensor chip is made of Silicon On Insulator (SOI) instead of the conventionally used Silicon. Fig. 1 demonstrates the superior high temperature potential of the dielectrically SOI-isolation compared to the p-n-junction normally used in silicon technology for piezoresistor isolation.

Due to leakage currents through the substrate the maximum operation temperature of silicon devices is limited to about 150°C, whereas SOI can be used at least up to 350°C.

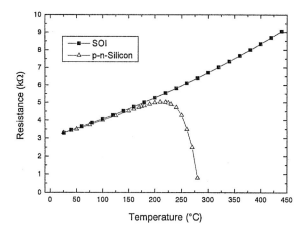

Fig. 1. Resistance of SOI and Silicon vs. Temperature

2.2 Sensor Fabrication

Using the characterised SOI material system, micromechanical piezoresistive sensor chips have been fabricated. Fig. 2 (left) shows a photograph of the sensor chip with the four piezoresistors that are switched in a Wheatstone bridge.

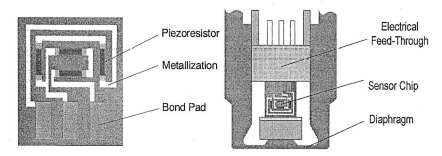

Fig. 2. Photograph of the SOI Sensor Chip (left) and Assembly of the SOI pressure Sensor (right)

The p-type piezoresistors are oriented in the <110> -crystal direction because this is the direction of the largest gauge factor. They are in pairs sensitive to longitudinal and transverse strains.

To be operated in combustion engines, the sensor chips need to be mounted into a specially designed housing. Fig. 2 (right) shows the assembly of the sensor with a M10 mounting thread and a robust steel diaphragm in front. No liquid is used for transferring the pressure induced deflection of the diaphragm to the wire-bonded sensor chip.

2.3 Sensor Characterization

During the measurement, both temperature and pressure are recorded from the sensor by measuring the bridge resistance as well as the output signal (Fig. 3).

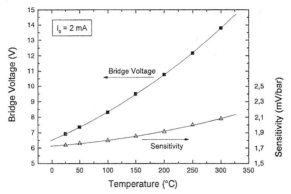

Fig. 3. Bridge Voltage and Sensitivity of the SOI Sensor Measured at Constant Current Supply

The bridge resistance at room temperature amounts to about 3.5 kΩ with a TCR of $+3 \cdot 10^{-3}$ K^{-1}. The sensitivity is ca. 1,7 mV/bar with a TCS of $+1 \cdot 10^{-3}$ K^{-1}. The temperature dependence and non-linearity of the pressure signal are compensated via an external electronic unit.

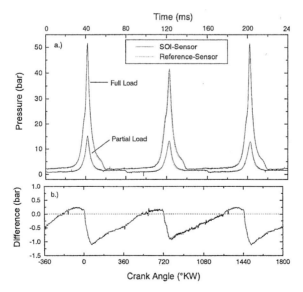

Fig. 4. Cylinder Pressure (a.) and Short-term Drift (b.) Measured with the SOI-Sensor

Fig. 4 shows the pressure signal measured in an internal combustion engine operated at 1500 rpm under full and partial load, respectively. A high-precision quartz pressure transducer from Kistler was used as a reference. The lower graph shows the short-term drift of the SOI-sensor monitored at full load operation. The maximum difference between the SOI sensor and the reference sensor amounts to about 1 bar.

The SOI-sensor can also be used for carrying out a gas exchange analysis and mass airflow estimation. Fig. 5 shows the low-pressure part of the cylinder pressure measured with the SOI-sensor in the combustion chamber compared to the signals monitored with piezoresistive reference sensors placed in the inlet and exhaust manifold. The combustion engine is operated at 1500 rpm and partial load.

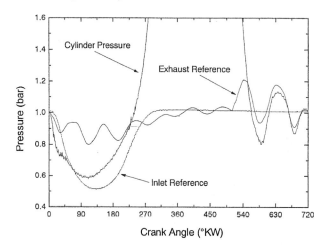

Fig. 5. Gas Exchange Measured with the SOI-sensor in Comparison to Reference Sensors Mounted in the Inlet and Exhaust Manifold

The most important sensor specifications are summarised in Table 1.

Pressure Range:	0...100 bar
Overload:	250 bar
Temperature Range:	-50...350°C
Mounting Thread:	M10
Sensitivity:	100 mV/bar
Natural Frequency:	50 kHz
Linearity Error:	< ±1 %FSO
Thermal Offset Shift:	< ±5 %FSO
Thermal Sensitivity Shift:	< ±1 %FSO
Thermo Shock Error:	< 1,5 bar

Table 1. Sensor Specifications

3.2.5 Benefits of a Cylinder Pressure Based Engine Management System

3 Control Algorithms

A number of model-based control and diagnostics algorithms were considered. These are: Spark timing control, Mass airflow estimation, Start control, Misfire detection, Knock detection, Cylinder 1 detection. These algorithms were developed in the MatrixX/SystemBuild simulation environment. The mass airflow estimation, cylinder 1 detection and knock detection algorithms can replace a number of conventional sensors thus resulting in further cost benefits.

- Spark Timing Control : The spark timing controller utilises pressure signals as a feedback variable thus allowing individual cylinder control. Spark timing can be controlled to maximise fuel efficiency or to achieve a smooth running engine.
  - Spark Timing Control for Near MBT: Maximum fuel efficiency can be achieved by controlling each cylinder to near Minimum advance for Best Torque (MBT). The Crank Angle (CA) at which 50% of the fuel is burnt is estimated at each cycle for each cylinder. This so-called 50% Mass Fraction Burn (MFB) point is controlled to a target CA for each cylinder. This provides robustness against engine production build variations as well as variations in humidity and fuel type.
  - Spark Timing Control for Smooth Running: Individual spark timing control also enables a smoother running of the engine. This smooth running mode is in contrast to achieving near MBT for each cylinder.
- Mass Airflow Estimation: Mass airflow was obtained by estimating the trapped air mass in each cylinder. This estimation algorithm is based on physical principles and includes the estimation of internal Exhaust Gas Recirculation (EGR) due to valve overlap.
- Start Control: Feedback control of individual cylinders enables better control during the warm-up phase of the engine. This allows greater spark retard enabling the catalyst to rapidly attain its optimal conditions (i.e. reduced catalyst light-off).
- Misfire Detection: Direct monitoring of cylinder pressures enables reliable detection of misfire. This facilitates an OBD function as well as providing essential information to the other feedback control and estimation algorithms.
- Knock Detection: In-cylinder pressure sensing provides an effective means for detecting cylinder knock. The time-synchronous pressure signal is filtered to extract the knock frequency component. Knock is detected when the amplitude of this component exceeds a given

threshold. Because of hardware constraints, the knock detection could not be tested on the prototype vehicle.

Cylinder 1 Detection: Cylinder 1 detection enables the cam phasing to be determined. This is made possible simply by comparing pressure differences in the compression/expansion phase with those obtained during the induction/exhaust phase. This provides a direct means of detecting the phase of the engine cycle. Again, because of hardware constraints, this strategy could not be tested on the prototype vehicle, but it has been tested in simulation.

4 Results

The prototype vehicle is an E-Class Mercedes-Benz with an M112 SI engine. This 3.2 litre V6 18 valve dual-spark ignition engine was installed with one pressure sensor per cylinder. Control and diagnostics algorithms were implemented on a rapid prototyping system.

This section presents some of the results of the CPEMS controller running on the prototype vehicle. It should be noted that the sample period of the data logging equipment was fixed at 0.8 seconds.

4.1 Spark-timing Control

Fig. 6 shows the results of the controller for near MBT timing. The controller was first running in open loop, as in a conventional EMS, until 42 seconds at which closed loop control began. The 50% MFB point for each of the six cylinder is seen to converge on to the target CA of 13° After Top Dead Centre (ATDC). Furthermore, tighter control of the 50% MFB points can be observed. The spread of spark timings across the cylinders during closed loop control provides an indication of the cylinder-to-cylinder differences that are compensated for by this feedback controller. These results were obtained at a constant engine speed of 2300 rev/min with the vehicle standing.

The near MBT strategy results in the optimisation of the combustion efficiency in all the cylinders. The global fuel efficiency is thus improved. The Indicated Specific Fuel Consumption (ISFC) obtained from estimated IMEP and fuelling commands has been compared between baseline strategy and CPEMS strategy. The results of steady static tests on vehicle indicate an average reduction in fuel consumption of 1.4%.

3.2.5 Benefits of a Cylinder Pressure Based Engine Management System

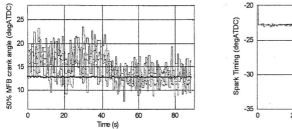
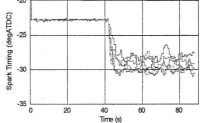

Fig. 6. Results of Spark Timing Control for Near MBT

The results of the spark timing controller operating in the smooth running mode can be seen in Fig. 7. In this example, the controller was first operating in the near MBT mode. Then at around 30 seconds, the controller was switched to the smooth running mode. This plot shows the results of the estimated IMEP after low-pass filtering. As shown, the initial spread in this IMEP across the cylinders of around 8% is reduced to within 2%.

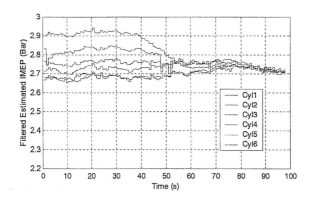

Fig. 7. Results of Spark Timing Control for Smooth Running Engine

The response of the controller to a transient load application is demonstrated in Fig. 8. A throttle tip-in was applied with the vehicle standing at 1800 rev/min. The smooth running algorithm monitors the cycle-to-cycle variations in IMEP. This enables transient load changes to be detected at which point the feedback controller is switched to near MBT timing on all cylinders in order to maximise engine torque for improved driveability. On subsequent detection of steady-state conditions, control is switched back to smooth running mode.

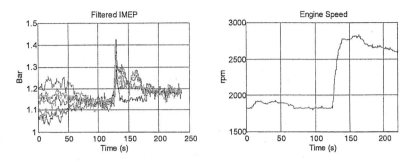

Fig. 8. Transition between Near MBT Timing and Smooth Running Mode

4.2 Start Control

The results of the start control are shown on Fig. 9. Both banks pre and post catalyst temperatures are displayed for an engine start using the production ECU strategies, and for an engine start using the CPEMS start control strategy. The pre-catalyst temperatures are 200°C higher with the CPEMS strategy than with the production one. This results in the reduction of the light-off time by 10% implying a significant improvement in emissions, and a possible reduction in the amount of precious metals required in catalytic converters.

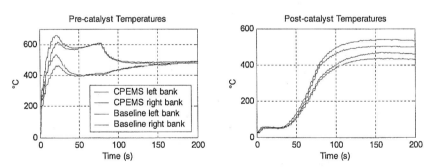

Fig. 9. Catalytic Temperatures with and without CPEMS Start Control Strategies

5 Conclusion

This paper describes a project aimed at demonstrating the benefits of CPEMS technology. The AENEAS project involved the combination of innovative sensing technology and intelligent control algorithms that will lead to a cost-effective solution to cylinder based engine management. A static piezoresistive pressure sensor based on a micromechanical SOI sensing element has been developed. The sensor is designed for a maximum pressure of 200 bar and operation temperatures up to 350°C. A number of algorithms have also been developed, some of which can replace conventional sensors. The results shown in this paper indicate that spark timing can be controlled to near MBT for maximum fuel efficiency or for a smoother running engine. In both cases, individual cylinder feedback control is shown to provide robustness against cylinder-to-cylinder variations and potential variations in spark timing caused by changes in fuel properties or humidity. Better control of the spark timing and consequently of the engine stability enables higher retard of the ignition during cold start of the engine. The reduction of catalyst light-off time by 10% will lead to improvements in emissions and a significant reduction in catalyst material.

Acknowledgements

This work was financially supported by the European Commission DGXII/D as Innovation Programme IN3010561 and the Swiss Government (Bundesamt Fuer Bildung Und Wissenschaft). The permission of DaimlerChrysler, EADS, Kistler and Ricardo to publish this paper is also acknowledged.

Special thanks are due to Marc Vigar, the Project Manager, and the rest of the engineering team. This included Chris Evans at Ricardo, and Robert Sindlinger and Martin Mueller at DaimlerChrysler.

The work described in this document is protected under the patent application GB0112338.9 21 May 2001.

References

[1] R. Mueller, M. Hart, A. Truscott, A. Noble, G. Kroetz, M. Eickhoff, C. Cavalloni, M. Gnielka, Combustion Pressure Based Engine Management System, SAE 2000-01-0928, (2000).

[2] J. von Berg, R. Ziermann, E. Obermeier, M. Eickhoff, G. Kroetz, U. Thoma, C. Cavalloni, and J.P. Nendza, 'Measurement of the Cylinder Pressure in Combustion Engines with a Piezoresistive β-SiC-on-SOI Pressure Sensor', 4th Int. HiTEC '98, Albuquerque, New Mexico, USA, pp. 245-249.

[3] J. von Berg, M. Gnielka, C. Cavalloni, T. Diepold, B. Mukhopadhyay, and E. Obermeier, 'A Piezoresistive Low-Pressure Sensor Fabricated Using SOI For Harsh Environment Applications', Transducers'2001, pp. 482-485.

[4] Michael Mladek; Christopher H Onder, 'A Model for the Estimation of Inducted Air Mass and the Residual Gas Fraction using Cylinder Pressure Measurements. SAE 2000-01-0958.

3.3 Measurement - Angle and Position Sensors for Vehicles

3.3.1 TPO (True Power On) Active Camshaft Sensor for Low Emission Regulation

G. Carta

Infineon Technologies AG
Balanstrasse 73, Munich, 81609 Germany
Phone: +49/89/234/21537, Fax: +49/89/234/21237
E-mail: giovanni.carta@infineon.com, URL: www.infineon.com

Keywords: camshaft, hall sensor, true power on, twisted independent mounting

Abstract

This "new member" of Infineon Technologies Sensor family, TLE4980, is an active chopped mono cell Hall sensor ideally suited to back-bias reduced camshaft applications. Its basic function is to map either a tooth or a notch into a unique electrical output state. The mono cell concept allows TIM, Twisted Independent Mounting.

The device is customer programmable in order to achieve TPO, True Power On capability (that means the correct output state is given instantaneously on switching the supply voltage on), even in the case of wide temperature changes and production spreads such as different magnetic configurations or misalignment. An additional self-calibration module has been implemented to achieve optimum accuracy during normal running operation conditions.

1 Introduction

Regulatory demands in Europe (EURO IV by 2005) and North America to restrict emissions during engine start-up require system suppliers to improve the efficiency of their "quick start" strategies. Cam shaft position accuracy plays a key role in this.

It is this aim that guided Infineon Technologies in the development of an Active Camshaft Sensor. Traditionally passive devices, mainly using the resistive technology. According to SA (Strategic Analytics) active sensors will represent 78% of the total sensor market in 2007.

Infineon finds that its Hall-Effect technology represent the best trade-off between performance, dimensions and cost. The advantages of active sensors are the improved accuracy and reliability, smaller dimensions and improved EMC performance. These features had been driving the development since the introduction of the first generation active sensors (several years ago). These goals, achieved with the TLE4980, provide the car manufacturers with a 2nd generation active sensor that allows them to fulfil the future severe environmental regulation. This has been possible with the implementation of smart functions such as dynamical self-calibrating algorithms, TPO capability and "end of line" programmability.

2 Functional Description

The basic operational mode of the TLE4980 map a „high" magnetic field (tooth) to a "low" electrical output signal and a „low" magnetic field (notch) to a "high" electrical output. A magnetic field is considered as "high" if the North Pole of a magnet shows towards the rear side of the IC housing.

For understanding the operation of the TLE4980 three different modes have to be considered: Initial mode, Pre-calibrated mode and Calibrated mode.

2.1 Initial Mode

The initial mode is activated immediately after power-up. The magnetic information is derived from a spinning Hall probe and chopped amplifier. The threshold information [B_{TPO}] comes from the PROM-register. The open collector output will be turned on or off by comparing the magnetic field against the pre-programmed value [B_{TPO}].

The B_{TPO}, which is an absolute magnetic value (e.g. 30 mT), is calculated and programmed by the customer, (see par. 5) to assure the switching under all environmental and air-gap conditions.

Once the IC has accurate information (typically after 2-3 teeth) on the absolute Max. and Min. magnetic fields, caused by tooth and notch respectively, it then moves into the pre-calibrated mode.

2.2 Pre-calibrated Mode

In the pre-calibrated mode the IC adjusts the system offset in such a way, that the switching occurs at the desired threshold level [B_{cal}]. The optimum threshold level, for a specific magnetic configuration (e.g. target wheel profile, back bias magnet strength, operating air-gap, etc.) is the one that provides the best phase accuracy despite the air-gap variation. To self-compensate environmental variation the B_{cal} is a relative value; as shown in Fig. 1 it can range, depending on k_0 value [0..1], from B_{Min} to B_{Max}. Thanks to the programmability of k_0-factor the device can easily adapt to different magnetic configurations always providing the best possible accuracy.

The threshold adjustment is limited to increments of approx. 1,5mT per calibration in order to avoid spurious information originating from large signal disturbances (e.g. EMC-events or similar).

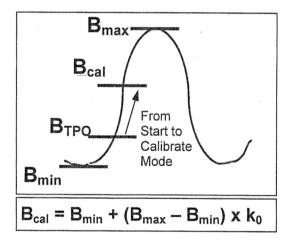

Fig. 1. Definition B_{cal}, Dynamic Switching Value

2.3 Calibrated Mode

After the B_{cal} level has been reached the IC moves into calibrated mode where only minor threshold corrections are allowed. In this mode a period of 32 switching events is taken to find the absolute minimum and maximum within this period. Threshold calculation is done with these minimum and maximum. A filter-algorithm is incorporated, which ensures that the threshold will only be updated, if the adjustment value calculated shows a change in the same direction over the last four consecutive periods. Additionally an activation level is implemented, allowing the threshold to be adjusted only if a certain amount (normally bigger than 1

LSB) of adjustment is calculated. The threshold correction per cycle is limited to 1 LSB. The purpose of this function is to avoid larger offset deviations due to sporadic singular events. Also irregularities of the target wheel are cancelled out, since the minimum and maximum values are derived over at least one full revolution of the wheel. The output switching is done at the threshold level without visible hysteresis in order to achieve maximum accuracy.

3 Hysteresis Concept

There are two different hysteresis concepts implemented in the IC.

The first one is called visible hysteresis, meaning that the output switching levels are changed between two distinct values (depending on the direction of the magnetic field during a switching event), whenever a certain amount of the magnetic field has been passing through after the last switching event. The visible hysteresis is used in the initial mode.

The second form of hysteresis is called hidden hysteresis used in pre-calibrated and calibrated mode. This means, that it is not possible to observe a hysteresis from outside.

4 Block Diagram

The block diagram is shown in Fig.2. The IC consists of a spinning-Hall-probe (mono-cell in the centre of the chip) with a chopped preamplifier. Next there is a summing node for threshold level adjustment. The threshold switching is actually done in the main comparator at a signal level of „0". This means, that the whole signal is shifted by this summing node in such a way, that the desired switching level occurs at zero. This adjusted signal is fed into an A/D-converter. The converter feeds a digital calibration logic. This logic monitors the digitised signal by looking for minimum and maximum values and also calculates correction values for threshold adjustment. The static switching level is simply done by fetching a digital value out of a PROM. The dynamic switching level is done by calculating a weighted average of min and max value. For accurate phase stability at different air gaps the dynamic switching level can be programmed with a weighting factor (k_0). For example, a weighting factor of approximately 67% can be achieved by doubling the weight of the max value. Generally speaking, a threshold level of $B_{cal} = B_{min} + (B_{max} - B_{min}) * k_0$ can be achieved by multiplying max with the switching level k_0 and min with $(1-k_0)$.

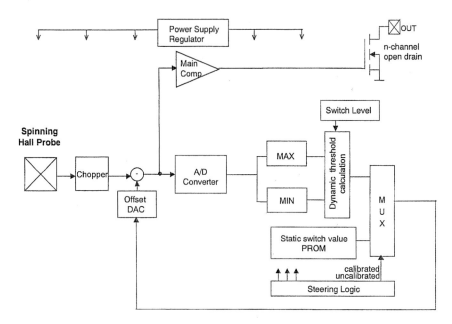

Fig.2. Block Diagram

5 Programming of B_{TPO} and K_0-factor

The programmability of B_{TPO} and k_0 is one of the major features of this device. As mentioned in par. 2.2 the ability to choose k_0 allows the device to cover different magnetic configurations without compromising accuracy.
The advantage of being able to program B_{TPO} when the device is already integrated in the final module or even fixed in the engine is that all the assembly imperfections and mechanical tolerances can be completely eliminated. There are basically two possibilities for programming the static threshold value B_{TPO}.

The first one is to run the IC on a test-bench (or in the engine directly), to wait until the IC has reached the calibrated mode and then simply to issue the copy commands, which transfers the calibrated threshold value into the PROM. A second way is to bring the IC in front of a target, which delivers a static magnetic field with a suitable strength and to perform a power-on by forcing the output to a low state for at least 1ms. This brings the chip in the test-mode; it starts immediately a successive approximation and adjusts the value of the offset-DAC to the switching level that corresponds to the field strength. At the same time the k_0-factor is programmed too. Programming is performed by a serial interface; the same interface can be used to provide special settings and to read out several internal registers status bits.

6 Package

TLE4980 is provided in an extra slim style [1 mm] plastic package [PSSO3-9]. Two 4.7 nF capacitors have been integrated in the lead-frame. This new configuration make the assembly much simpler, and hence also cheaper, for the customer as no PC board is required to mount discrete external components. The two capacitors have been optimised to achieve excellent EMC performance levels.

Fig. 3. TLE4980 TPO Active Camshaft Sensor

7 Summary and Outlook

The development of the Infineon TLE4980 marks the introduction of a new generation of Active Cam-shaft Sensor technology. Thanks to the wide use of smart functions such as advanced dynamical algorithm, TPO capability and programmability, the sensor works in perfect symbiosis with the mechanical and magnetic components of the system. The "end of line" programmability at customer package level, in fact, allows a tailoring of each sensor to the engine onto which it is fixed for its operating lifetime. Infineon, coherently with the policy of continuous improvement, is implementing these features also to the sensors for crankshaft and transmission applications.

References

[1] "Active Sensor Market Trends, Part I", Strategic Analytics, January 2001.

3.3.2 Contactless Position Sensors for Safety-critical Applications

Y. Dordet[1], B. Legrand[1,2], J.-Y. Voyant[2], J.-P. Yonnet[2]

[1]Siemens VDO Automotive S.A.S., Sensors Innovation
BP 1149, Avenue Paul Ourliac, 31036 Toulouse Cedex, France
Phone: +33/5/61/197294, Fax: +33/5/61/192535
E-mail: yves.dordet@at.siemens.fr

[2]Laboratoire d'Électrotechnique de Grenoble
ENSIEG, BP 46, 38402 Saint-Martin-d'Hères Cedex, France

Keywords: contactless, magnetic saturation, printed coil, position sensor, reliability, safety, standardised technology

Abstract

In the coming decade, many functions in the cars will be managed by wire (i.e. drive-by-wire, steer-by-wire, brake-by wire). It means that the actuating systems will increase the number of position sensors. Today the resistive potentiometers are the most frequently used sensors. For reasons of reliability, there are increasing demands to replace these kinds of sensors by contactless displacement sensors, with almost no increases in the system costs.

The use of contactless sensor could be considered for safety-critical application only if the reliability allows a complete trust in the delivered information.

Beside the "contactless" demand, the sensor should work without friction of the moving parts and thus allowing big air gaps.

A high performance transducer technology for displacement sensing coupled with the functionality of an useful ASIC has been developed to fulfil these requirements.

1 Introduction

The transducer design is based on a PCB (Printed Circuit Board). The technology of the PCB is a standard one, rigid or flexible, that permits to get the sensitive area at a competitive cost. At least one layer is used to print a coil in the shape of the required sensor type. This coil is covered by a structure with highly permeable magnetic material. The proximity of this

3.3.2 Contactless Position Sensors for Safety-critical Applications

material increases the value of the printed coil inductance. A magnet in the moving target creates, in the ferromagnetic material, a saturation area proportional to its position and so it changes the inductance of the printed coil.

A dedicated ASIC is designed to manage the measurement of the inductance change and to insure the communication with the electronic control unit (ECU). For several application the transducer and its dedicated electronics can be mounted directly on the PCB of the ECU.

All kinds of displacement sensors (linear, short or long travel, rotary, and even 2 dimensional) are feasible in a very flat design.

Due to the saturation effect, the sensor has a very low sensitivity to airgap changes and even allows airgap of up to 6mm, even through non ferromagnetic walls !

Planar-coil displacement sensors fulfil automotive specifications and in comparison to resistive potentiometers, offer many advantages such as: no wear, easy assembly, no pollution risks, large airgaps, etc...

The electronic management and communication protocol allow the use of these sensors in critical functions in the car. They will be used for robotized transmissions, clutch actuators, Electro-Hydraulic Brake, steering wheel position sensing, throttle and other valve position sensors...

2 Sensor Model

The sensor design is clarified with the models of the sensors (linear and rotary) for finite-element simulations. We used the well-known FLUX3D package.

2.1 The Sensor Design

The system works on a differential mode in order to minimise external disturbances including airgap changes or mechanical tolerances, offset errors and other non-linearities. This differential mode is also used during the finite element simulations.

The design of the rotary sensor is as follows (Fig. 1):

- Two rectangular coils made of 2 layers on a PCB. The two layers are interconnected using vias (through hole plating) and are wound in the same direction in order to increase the value of the inductance.

- A soft magnetic sheet with a high permeability (60000) and a very low coercitivity (1 A/m, 12,5 mOe, low hysteresis). This sheet is the easiest path for the magnetic field generated by the coils. Actually Mumetal is used improving the reliability of this technology.

A movable magnet (rare earth magnet or plastic bonded rare earth magnet) creates a saturated area on the magnetic sheet and reduces the magnetic field from the coil. The value of the inductance depends on the magnet position.

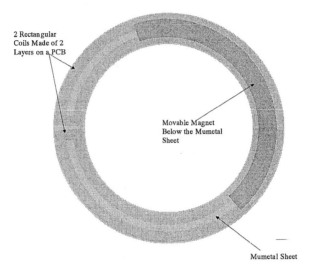

Fig. 1. Rotary Sensor Design

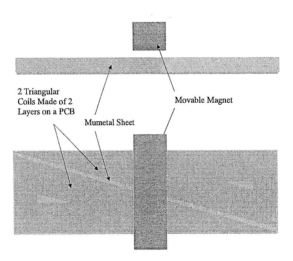

Fig. 2. Linear Sensor Design

The design for the linear sensor contains two triangular coils (for differential mode). This allows to reduce the size of the magnet (Fig. 2).

2.2 Finite Element Model

Using the finite-element software Flux3d, the geometry of the sensitive element and the magnet could be optimised. It permits to reduce the magnet size, to simulate the sensor behaviour depending on the customer mechanical specification (airgap change, lateral or angular shift, magnet strength).

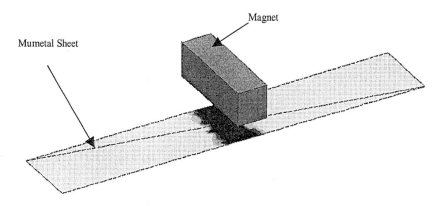

Fig. 3. General Overview of the Saturation for the Linear Sensor

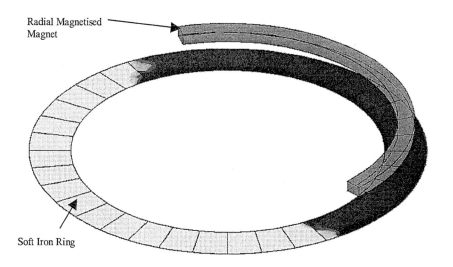

Fig. 4. General Overview of the Saturation for the Rotary Sensor

Fig. 3 and 4 present the saturation for a linear and a rotary sensor. The yellow colour corresponds to the non saturated area, the blue for the saturated area (at 0,8 T) created by the magnet and the green for the magnet.

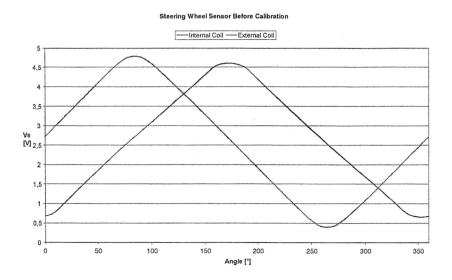

Fig. 5. Measured Output Signal for a 360° Rotary Sensor

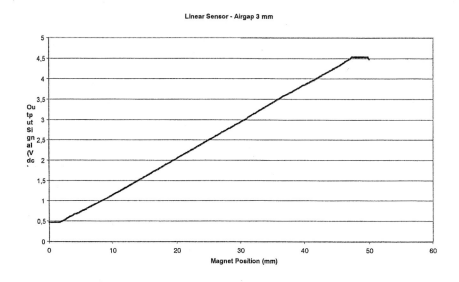

Fig. 6. Measured Output Signal for a Linear Sensor

Fig. 5 and 6 represent the real output for a rotary and a linear sensor. For the rotary sensor the Mumetal sheet is circular and does not contain edge. On the contrary, for the linear sensor, the Mumetal sheet is rectangular

3.3.2 Contactless Position Sensors for Safety-critical Applications

and thus contains two edges where saturated area and magnet movements are different, which creates a non-linearity in the output signal. For linear sensors the coils have to be longer than the required linear stroke.

The global dimensions of a sensitive element depend directly on the airgap and the stroke.

3 Signal Treatment

The coils are excited with a rectangular voltage signal. After synchronous rectifying and integration of the signal in the middle of the half-bridge, the signal is ready for further digital signal processing in a micro-controller (Fig. 7).

For stand alone sensors, an ASIC (Application Specific Integrated Circuit) has been developed where analog and digital signal processing are on one IC.

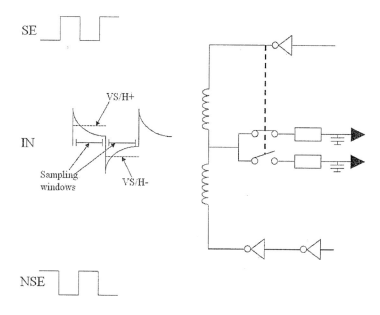

Fig. 7. Sensor Excitation

3.1 Sensor Output

There are 2 outputs:

- a ratiometric analog output (10% to 90% of V_{in}) with a resolution of 10 bits and a data rate of 5 kHz.
- a digital output using LIN protocol (Local Interconnection Network) with a resolution of 9 bits and a data rate of 0,5 kHz.

A disconnection of one output does not affect the second one.

An information of the sensitive area temperature is available on the LIN sentence, coded on 6 bits. This information could be used to prevent sensor or system malfunction due to an occasional use outside of its temperature range. For example the start of the motor at a temperature less than - 40°C or an abnormal temperature in a gearbox.

3.2 Sensor Calibration

The digital signal conditioning contains a programmable calibration of gain, offset and temperature drift coefficient. For each sensor, parameters are set on a test bench at the end of the production line. An EEPROM contains all these values. This component is designed to insure a safe function even after 10 years at a continuous temperature of 150°C. During the normal sensor operation these parameters are transferred in a shadow register. Losing data in the EEPROM will never affect the sensor and will be detected during the sensor start up.

The data set in the EEPROM could be read and programmed using the LIN or the analog pin. A special voltage switches the analog output in a dialog interface with the ASIC and EEPROM. If needed this data could be definitively locked preventing a future change.

3.3 Sensor Diagnosis

The ASIC performs temperature measurement and diagnosis. The temperature is used for temperature drift compensation using the parameters programmed in the EEPROM and dedicated to the application.

The diagnosis is done detecting a wrong impedance of the sensing element. When a default occurs, the output is settled in tristate. The sensor recovers normal functioning as soon as a good new impedance is measured again. The diagnosis result is given by the analog output (5 V with pull-up) and by the LIN (one bit of the sequence especially dedicated). The electronic circuit is designed to give a tristate (Gnd, V_{dd} or infinite impedance) in case of bad connection or short circuit between two connecting pins. By no means an abnormal signal could be assimilated as a working position.

3.4 Sensor Accuracy

Accuracy is dependent on the real sensor configuration. Important parameters are sensor diameter, airgap changes, coaxiality and temperature range.

To compensate the customer interface and the mechanical tolerances of the system it is always better to reprogram the gain and offset of the sensor on the customer test bench or directly in the car if needed. Using this possibility, the final accuracy takes into account the real environment of the sensor like magnet strength, angular and shift error of the sensor position.

In fact this technology allows large tolerances of the magnet position in relation of the sensor. The typical accuracy of theses sensors is less than ± 2% over the temperature range of - 40°C <=> + 140°C.

4 Applications

The working principle described above can be incorporated into many kinds of rotary and linear displacement sensors and even 2 dimensional sensors as a standardised technology.

The unique point is the big airgap and thus the possibility to detect the position of a target through metallic but non-ferromagnetic walls.

In more and more cases the sensor is not a stand-alone sensor, but is integrated in the control device of the application.

Some examples of applications:

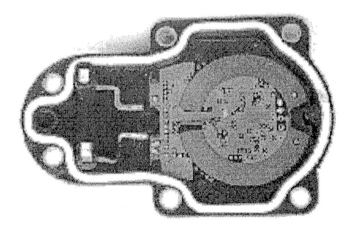

Fig. 8. Valve Actuator with 100° Rotary Sensor

Fig. 9. Linear Displacement Sensor 35 mm for AMT(Automated Manual Transmission), EHB (Electro-Hydraulic Brake), etc...

Fig. 10. Rotary Sensor 360° Integrated for Steering Column Module

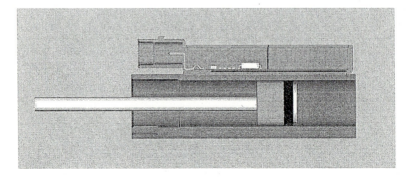

Fig. 11. Linear Displacement Typical Application

5 Conclusion

The described sensor technology offers the system designers a new freedom for advanced product conception. It can be integrated in a small volume and easily adapted to any geometry, angular or linear. The use of a standard technology and material makes the design cost effective.

With this large range of use, IMS (Inductive Magnetic Saturation) can become the standard technology for displacement sensing.

Integration into the PCB allows a better system integration and opens the way to cost effective "SMART"-devices for mechatronics.

The functionality of the associated electronic (calibration, diagnosis, digital and analog output...) open the way to safety-critical use of contactless sensors in the automotive industry.

Acknowledgments

The development of this technology is done in co-operation with the Laboratoire d'Électrotechnique de Grenoble.

References

[1] W. Gopel, J. Hess, J.N. Zeme, "Sensors a comprehensive survey", Vol.5 Magnetic sensors, VCH, Weinheim, 1989, p. 269.

[2] G. Asch et Coll., "Les capteurs en instrumentation industrielle", Dunod, Paris, 1998, pp. 319-394.

[3] E. Du Trémolet de Lacheisserie, "Magnétisme", PUG, Grenoble, 1999, pp. 93-189.

[4] "Conditionneur de signaux pour transformateur à variation linéaire", Electronique, N°96, October 1999.

[5] R.H. Darling, "High Reliable position sensors", IBM. Techn. Bull., Vol.12, N°4, Sept.1969, p. 536.

[6] D. Holt, "Sensors and automobile", Automotive Engineering International, vol 106, n° 9, Sept. 1998.

[7] W.F. Grover, "Inductances calculations", Dover Edition, New-York, 1962.

3.3.3 High Resolution Absolute Angular Position Detection with Single Chip Capability

H. Grueger, H. Lakner

Fraunhofer Institut für Mikroelektronische Schaltungen und Systeme
Grenzstrasse 28, 01109 Dresden, Germany
Phone: +49/351/8823-155, Fax: +49/351/8823-266
E-mail: heinrich.grueger@imsdd.fhg.de, URL: www.ims.fhg.de

Keywords: angular position detection, high resolution, magnetism, hall sensors, single chip

Abstract

A linear array of hall sensors has been realized in a 100% CMOS compatible technology. A cylindrical magnet, diametrally magnetized, mounted at the middle of the turning axis provides a pattern, that has a zero field position depending on the angular position.

Hall array and magnet form a system, that empossibles angular position detection using a linear regression over the signals of some of the hall sensors. Applying a linear regression over the signals of eight sensors a resolution below 0.1° is possible. The position value is available at once after power up. Due to single standard CMOS chip capability, a reasonable price is achievable.

1 Introduction

In many applications the detection of an angular position is required. Several different types can be necessary. For systems like throttle valve controls the absolute position within a limited range, like 0..90°, will be sufficient. In the field of rotation axes (crankshafts, camshafts,...) or some other applications like steer by wire the full circle (0..360°) has to be detected. Further requirements beneath the absolute position detection are long time reliability, a high resolution and immediate position values after power up. The invention in a higher volume market like the automotive sector is price sensitive. A single standard CMOS chip with reasonable area, mounted in a low cost technology on a printed circuit board (PCB) and an inexpensive signal generation system has important advantages.
Some years ago linear arrays of hall sensors have been developed for applications like linear position detection (1). The range of applications can

be extended to rotational movement, if a suitable magnet for the pattern generation is available. Advantages of this system are the possibility to integrate sensor, driving circuitry, signal processing, non volatile memory and interface in one single CMOS chip. The signal generation magnet can be small which implies a reasonable price even for rare earth magnets. The whole system is very stable, resistant to the automotive environment and maintenance free.

2 Realization

The electron movement correlated with an electrical current in a conductor results in a Lorenz force in a magnetic field perpendicular to the movement. This leads to a charge transfer which can be measured as the hall potential or voltage. In a known system the value of the hall voltage can be correlated with the magnetic field strength. Hall sensors can also be realized as silicon integrated devices. A crucial point is that in mono crystalline silicon additional voltages occur due to the piezo electric effect. One possibility to avoid disturbance of the magnetic field measurement is usage of so called "spinning current" hall sensors. A photograph of a single sensor is shown in Fig. 1.

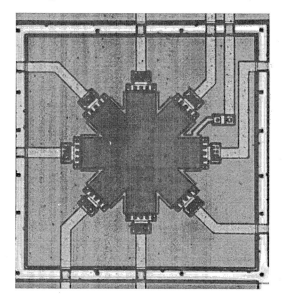

Fig. 1. Photograph of a Single Spinning Current Hall Sensor

The current is imprinted in changing directions, the measurement takes place in the perpendicular direction. After one or more sets of all directions possible (here: eight possibilities), the result is evaluated. Due to the symmetry, the piezo electric effect disturbance is compensated. Typical field values which can be measured using integrated hall sensors are in the millitesla [mT] range.

Using a linear array of hall sensors with a well defined distance, a length scale can be build. Every characteristic value of a linear moving magnet, for example the zero field point, can be detected with high accuracy. Further enhancement of the position detection can be achieved using a regression over the values of a couple of sensor in the surroundings of the zero field value. As only the neighboring sensors are regarded, these are the only which must not reach saturation. Thus it is possible to design the system for a steep inclination. With the same chip the detection of a rotational movement is possible if a correlation of the angle and the zero field position can be achieved. One possibility is a cylindrical magnet, that is diametrically magnetized. Using this, the angular position value is correlated with the position of the zero magnetic field value. For some applications this will be a sufficient solution, but for some tasks additional work is necessary to overcome the nonlinear behavior of the accuracy and an unambiguous signal over the full 360° range.

3 Experimental Results

Integrating 32 hall sensors with approximately 150 µm pitch a distance scale with over 4 mm range has been built (Fig. 2). The linear position of

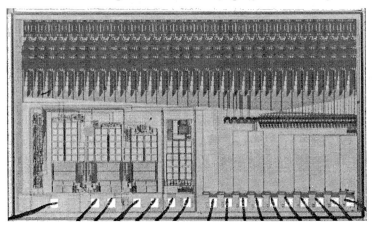

Fig. 2. Photograph of the Hall Array Chip with 32 Sensors with 151,2 µm Pitch

3.3.3 High Resolution Absolute Angular Position Detection

zero field value of a linear moving magnet was detected with an accuracy below 1μm. This was reached with a linear regression over the signal values of eight sensors. An output data rate up to 40 kHz was possible.

High field magnets with a diametric magnetization have been provided by Magnetfabrik Schramberg. Applying this magnet, the zero field values linear position depends on the rotational position if the magnet is turned and not moved. Using this the detection of the angular position is possible. Memorizing the pattern in non volatile memory the angular position can be detected with 0.1° resolution over a 90° range. For the further range the pattern has to be viewed more exactly. The linear movement of the zero field position under rotation has a sine form. That means, close to the turning points the movement is close to zero and the angular position detection has a far less resolution. This can be overcome using two magnets turned 90° towards each angular position. A second point is that for each pattern exactly two positions with the same one dimensional position exist. An additional one-bit information, which side of the magnet is viewed, is necessary. This can be reached through an additional pattern or through a double line array or other kinds of two dimensional hall arrays.

4 Discussion

The system described here has some favorable advantages. In a range up to 90° the resolution is quite high. The system has a high long time stability because changes in the magnetic field strength or sensitivity do not change the zero field position. Only the inclination of the regression and thus the accuracy of the position fit is affected, not the value itself. The system offers the possibility to integrate all components up to the interface into one standard CMOS chip. A simpler mounting technology than for the established analog sine/cosine evaluation system can be applied. With a magnetic working principle no affection through dirt like in optical systems will occur. Finally hall sensors can be produced with high temperature capability which makes them applicable close to engine parts.

In fairness some disadvantages should not be hidden:

Good accuracy of rotation axis required, longitudinal movement disturbs the position detection. Also parallel mounting of magnet and chip is necessary.

For the use over the whole 360° range some challenges have to be solved. This is the change in the accuracy close to the turning points and the appearance of the same pattern two times over a 360° rotation using a single linear array.

5 Conclusions

A system has been realized for absolute angular position detection. The resolution in a 90° range is at about 0.1°. By now additional one-bit information is required for an unambiguous correlation of the whole 360° range. The accuracy is 0.1° or better for a 90° range, but over the full circle the resolution and with it the accuracy is depending on the position.

6 Future

Two dimensional arrays or a double line of hall sensors will give further options. Maybe two magnets or a special magnetization will help to overcome the changes in the resolution and result in a constant accuracy over the whole range of operation.

Acknowledgements

The authors would like to thank Magnetfabrik Schramberg for providing the magnets.

References

[1] Gottfried-Gottfried, Ralf; Hauptmann, Dirk; Krause, Martin; Kreisig, Bernd; Ulbricht, Steffen: "Absolute position detection system with µm resolution using an integrated array of hall sensors" 9th Sensors conference publications p.261-262, 1999.

3.3.4 Programmable Linear Magnetic Hall-Effect Sensor with Excellent Accuracy

S. Reischl, U. Ausserlechner

Infineon Technologies AG
Balanstraße 73, 81541 Munich, Germany
Phone: +49/89/234-20345, Fax: +49/89/234-716418
E-mail: stefan.reischl@infineon.com, URL: www.infineon.com

Keywords: angle sensing, hall, linear, pedal position potentiometer replacement

Abstract

Infineon Technologies AG presents a linear output Hall-Effect sensor which specifically meets the demands of highly accurate rotation, angle and position detection, as well as current measurement applications. The sensor provides a ratiometric analog output which is ideally suited for A/D conversion using the supply voltage as a reference. The transfer function of the linear Hall IC TLE4990 can be adopted arbitrarily to the requirements of the application in terms of offset (zero field) voltage, sensitivity and clamping. Bipolar or unipolar mode can be selected to cover a wide range of automotive applications like pedal position or steering angle. The programming is implemented by using low current fusible poly-silicon links, which are accessed through a three wire interface.

Excellent accuracy is assured through the programmable temperature compensation of the sensitivity. Outstanding stability is achieved through dynamic offset cancellation to eliminate any parasitic mechanical or temperature effects. Produced in 0.5 μm BiCMOS technology the IC provides high voltage capability as well reverse polarity protection. A 4-pin plastic single in-line, small outline package of only 1.0mm thickness allows insertion into the smallest air gaps and is therefore optimising the available magnetic flux.

1 Introduction

Traditional potentiometers with sliding mechanical contacts are increasingly being replaced by contactless magnetic versions, wherever high reliability and long lifetime is of primary importance. The principal function of a potentiometer is to render an output voltage, which is proportional to the position of its slide and to the voltage across its fixed contacts (cf. Fig. 1). The same task can be fulfilled by a magnetic potentiometer, which provides an output voltage proportional to the magnetic flux density on the ASIC and to its supply voltage (cf. Fig. 2). The magnetic flux density changes according to the position of a lever, which carries a small permanent magnet. As with resistive potentiometers both linear and rotary position can be encoded with high precision. However, the magnetic potentiometer exhibits a number of additional advantages: (i) improved overall accuracy by in-circuit calibration after assembly of the module, (ii) less friction and wider range of velocities as there is no direct mechanical contact, (iii) larger operating temperature range and fewer limitations to humidity, dust, and chemicals (iv) improved repeatability due to the absence of mechanical hysteresis of the sliding contact, and (v) reduced hazard of mechanical damage resulting from limited travel.

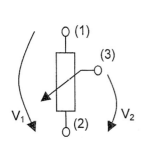

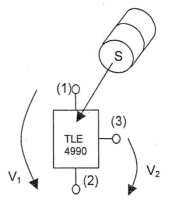

Fig. 1. A resistive potentiometer with sliding contact (3) gives an output voltage V2 proportional to the position of the slide and to the voltage V1 across its fixed contacts (1), (2).

Fig. 2. A ratiometric linear magnetic sensor can be configured as a magnetic potentiometer, whose output voltage V2 rises proportional to the supply voltage V1 across (1), (2) and proportional to the field of an approaching magnetic south pole.

Another field of application for linear magnetic sensors is current measurement. The electric current through a conductor can be measured indirectly by way of its magnetic field along a closed path around the conductor. The main advantages of measuring the current via its magnetic field are: (i) the power dissipation does not depend on the magnitude of the measured current, (ii) the measurement circuit has no Ohmic contact to the high current path, (iii) no insertion losses, i.e. the high current path does not have to be interrupted for insertion of a current shunt, and (vi) the linearity of magnetic sensors does not suffer from thermal effects as do resistive shunts. Current measurement without Ohmic contact is particularly suited for high current and /or high voltage applications.

2 Integrated Hall Sensor and Dynamic Offset Cancellation

2.1 Integrated Magnetic Hall Sensor ASICs

An integrated Silicon Hall-effect element is perfectly suited to versatile, high volume, and robust magnetic field sensing applications. Besides the magnitude of the magnetic field it can also detect its direction, and it does not suffer from the application of excessive field strengths (magnetic overload). Programmability allows to calibrate the ASIC individually during the customers manufacturing process, i.e. after mounting into the complete module. Thereby the tolerances of the magnet, the sensor and of the mechanical system can be compensated for. The sensitivity window of integrated Hall-effect elements ranges from several mT to hundreds of mT, depending on both the on-chip signal conditioning systems and on the system bandwidth. For these small field strengths the linearity of the Hall-effect element is practically perfect. This makes the Hall effect ideally suited for use in a linear magnetic sensor, which has a linear transfer function output voltage versus magnetic flux density.

2.2 The Concept of Dynamic Offset Cancellation

In the past there has been one major drawback of Hall-effect elements: their output signal does not vanish at zero magnetic fields. A major source of offset is mechanical stress which is exerted from the package onto the die and which is inhomogeneous, temperature dependent, hysteretic after thermal cycles, unstable over lifetime and practically not predictable [1], [2]. Despite perfectly symmetric geometry and technology, mechanical stress can lead to an asymmetry of the electric properties of the Hall-effect probe, due to piezo-resistive effects. This is the main reason why Hall-effect probes are often placed in the center of the die, where mechanical stress is most homogeneous. One strategy to reduce the stress-induced offset is known as orthogonal coupling of Hall-effect probes, where two Hall-effect probes are connected in parallel, however, one is rotated against the other by 90° in the layout [3]. Despite a substantial reduction down to a few mT by use of orthogonally coupled or quad-coupled Hall-effect probes the remaining equivalent magnetic offset field is still too large for most applications.

The main reason for residual offset is that the individual Hall-effect plates in an orthogonally-coupled Hall probe are neither perfectly identical nor do they experience the same mechanical stress. Therefore it is better to use a dynamic offset cancellation technique, where a single Hall-effect probe is connected in alternate clock cycles in the very same fashion as each plate of the orthogonally-coupled Hall-effect probe is connected. The time averaged samples of the Hall-effect voltage over one period is representing a Hall element signal, which is practically independent of the orientation of the Hall-effect element and is therefore independent of anisotropic effects. In addition its sensitivity to small asymmetries is also greatly reduced, as is shown in [4]. We call this strategy a 2-phase switched Hall-effect probe. It is similar to the well known technique of dynamic element matching.

3 System Architecture and Signal Processing

Fig. 3 shows the straight-forward block circuit diagram of the TLE4990. A modulator connects the signal of the 2-phase switched Hall-effect probe with proper polarity to a programmable gain differential-in : differential-out pre-amplification chain. The amplification factor is programmable by 3 bits (called PRE bits) with a dynamic range of roughly 1:11. This coarse gain adjustment selects the magnetic full scale range between 20 mT and 300 mT. The pre-amplification chain is followed by a demodulator and a re-settable integrator, which functions as sample-and-hold circuit. The demodulator changes the sign of the signal in every second clock phase, so that besides the Hall-effect element offset also the DC-offset and 1/f-noise components of the pre-amplification stages are effectively subtracted during the subsequent integration. The amplification of the integrator can be tuned by 10 bits (called GDAC bits, fine gain adjustment). Both coarse and fine gain adjustment cover the range of magnetic sensitivities from 15 mV/mT up to 235 mV/mT. A zero field voltage V_{zero} – programmable via 10 bits (called ODAC bits) – is added after the signal has passed all gain trimming stages, so that sensitivity S and zero field voltage can be adjusted independently.

3.3.4 Programmable Linear Magnetic Hall-Effect Sensor with Excellent Accuracy

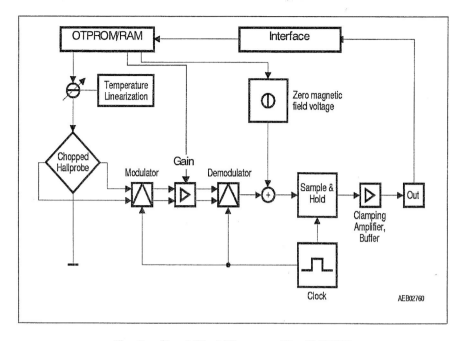

Fig. 3. Circuit Block Diagram of the TLE4990

Acknowledgements

We would like to especially thank H. Altrichter, W. Ebner, M. Holliber, W. Marbler, M. Motz, M. Mueller, H. Schatzmayr, G. Veranu, A. Zannantonio and R. Hipp for their support of this project. Part of this work were supported by the Austrian Industrial Research Promotion Fund (FFF), which is gratefully acknowledged.

References

[1] Y. Kanda, M. Migitaka, "Effect of mechanical stress on the offset voltage of Hall devices in Si IC", Phys. Status Solidi (a), vol 35, 1976

[2] Y. Kanda, M. Migitaka, "Design considerations for Hall devices in Si IC", Phys. Status Solidi (a), vol 38, 1976

[3] J. T. Maupin, M. L. Geske, "The Hall effect in silicon circuits", in The Hall Effect and its Applications, C. L.Chien and C. R. Westgate, Eds., New York, NY: Plenum, 1980

[4] P. Munter, "Spinning-current method for offset reduction in silicon Hall plates", thesis, Delft University Press, 1992.

3.4 Measurement - Pressure Sensing in Automotive Applications

3.4.1 Overview: Principles and Technologies for Pressure Sensors for Automotive Applications

C. Cavalloni, J. von Berg

Kistler Instrumente AG
Eulachstrasse 22, 8408 Winterthur, Switzerland

Keyword: pressure sensor

Abstract

This paper will give an overview on current and future pressure sensor principles and technologies for automotive applications. Microelectromechanical (MEMS) technology based on silicon has strongly influenced the development of pressure sensors and its applications. Such MEMS pressure sensors are of growing importance and a change from traditional discrete mechanical technologies to complete systems based on MEMS devices is in evolution. In special niches like high pressure and high temperature in the engine environment strain gage, piezoelectric or optical pressure sensors are the preferred solution.

1 Introduction

Pressure sensors are since more than half a century well established in automotive applications [1]. On the one hand, the development and testing of vehicles and engines requires robust and high performance pressure sensors, on the other hand, sensors and systems for pressure monitoring and control are needed in every commercial car. In the latter case the requirements for the sensors are high repeatability, insensitivity to the hostile environment and high reliability over the lifetime of a car, all achieved at significantly lower cost than equivalent industrial sensors. The driving force for incorporating more and more sensors is given by the requirements by legislation, decrease of emissions, increase of fuel efficiency, improvement of reliability, better comfort, higher safety and better handling. With the introduction of microsystem technologies high volume production of sensors and systems at low component cost has become possible. Microelectromechanical systems (MEMS) are now widely used for pressure sensing [2].

Typical pressure sensor applications for vehicles include monitoring of:
- Oil pressure
- Manifold absolute pressure (MAP)
- Barometric absolute pressure (BAP)
- Exhaust gas recirculation (EGR)
- Combustion pressure
- Common rail pressure
- Fuel injection pressure
- Hydraulic pressure in braking systems
- Pressure in air conditioning systems
- Tire pressure

The total value of pressure sensors represents about 10% of all automotive sensors. New developments for advanced engine management systems are including combustion pressure sensors in order to control the engine. In this case the information delivered by the pressure sensor is huge, allowing the replacement of different existing sensors. In the following sections different pressure sensor principles will be reviewed and its potential for automotive applications discussed.

2 Pressure Sensor Principles

Several different physical measuring principles and technologies have been developed for sensing pressure. However, only few of them have found wide applications in the automotive market. Mainly MEMS pressure sensors based on Silicon micromachining have made continuous progress towards improvement of performance, reliability and system flexibility. Resistive and capacitive sensors are the most used, piezoelectric sensors possess their well established niche, whereas optical, resonant or surface acoustic wave (SAW) sensors may become of increasing importance in the future.

Basically, the measurement of pressure is achieved by using a deformation element, typically a diaphragm, which separates two media (gas or fluid) and deflects when the pressure changes. This deflection may be measured directly, or through surface strains, or be transferred to a pressure or force acting on a sensing element. The conversion of the mechanical or displacement signal to an electrical signal is achieved in the sensing (transducer) element by making use of different physical principles (see Table 1):

- Electrical resistance change as function of strain
- Capacitance change as function of displacement
- Electrical charge generation as function of force
- Resonance frequency change as function of strain
- Optical quantity (light intensity, phase, etc.) change as function of displacement

Principle	Mechanical quantity	Electrical quantity	Sensing element
resistive strain gage	strain	electrical resistivity	strain gage, thin film
piezoresistive	strain	electrical resistivity	silicon, semiconductor, thick film
piezoelectric	stress	electrical charge	quartz, crystal, piezoceramic
capacitive	displacement	capacitance	ceramic capacitor, silicon structure
resonant	strain	frequency	resonator, silicon structure, SAW
optical	displacement strain	photocurrent	photodiode
inductive	displacement	inductive coupling	inductive coil, linear variable diff. transformer LVDT

Table 1. Pressure Sensor Principles

Of these pressure sensor principles the piezoresistive principle combined with MEMS devices has become the most favoured one for high volume production for automotive applications. In the following sections, the most important or promising measuring principles for automotive pressure sensor applications will be presented in more detail.

2.1 Resistive and Piezoresistive Sensors

Resistive strain gage and piezoresistive sensors are based on the change of electrical resistance as function of applied strain (or stress). The resistor element may consist of a metallic or piezoresistive (silicon) strain gage bonded to a stainless steel diaphragm, or a metallic thin film, or thick film deposited on the diaphragm (stainless steel or ceramic), or semiconducting piezoresistors incorporated directly in a micromachined silicon chip diaphragm. Normally the sensing diaphragm is exposed to the external media (Fig. 1).

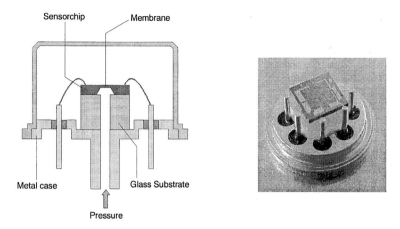

Fig. 1. Piezoresistive silicon pressure sensor with sensing diaphragm directly exposed to the external media.

However, in high temperature applications, or when silicon chips are used in harsh environments, a second diaphragm is used to separate the media. In this case, the pressure is transmitted through oil or a mechanical element to the sensing diaphragm (Fig. 2). In the latter (oil free) case, instead of the diaphragm another mechanical deformation element (e.g. a cantilever or beam) may be used as sensing element (see Fig. 3).

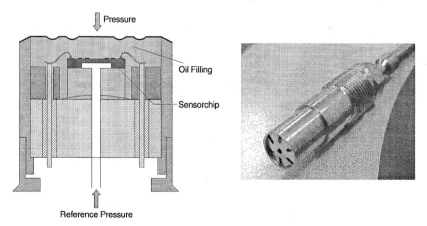

Fig. 2. Piezoresistive pressure sensor packaged inside a metal housing with oil filling.

Piezoresistive silicon pressure sensors are manufactured on wafer level in high volumes. They offer the possibility of integration of the electronic circuit on the same chip (MEMS). The maximum temperature for piezoresistive silicon pressure sensors is limited to about 150°C. With new silicon on insulator (SOI) technology the maximum operating temperature may be increased to about 350°C (Fig. 3).

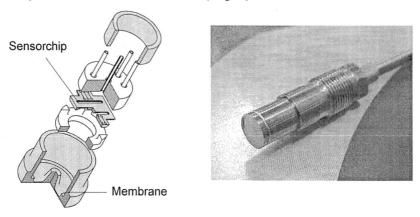

Fig. 3. Piezoresistive pressure sensor with front membrane and a sensor chip made of SOI.

A typical high temperature sensor application is found in measuring the pressure in the combustion chamber of an engine [3] or in the exhaust outlet [4]. Here, the pressure induced deflection of the steel diaphragm is transferred directly by mechanical elements on the sensor chip. Packaging and interconnection technology for harsh automotive environments represent the major challenge for silicon sensors. New piezoresistive materials such as silicon carbide or gallium nitride are still in the research and technology development stage. They are attractive for high temperature applications above the SOI temperature range.

Fig. 4. Silicon sensor chips fabricated using high volume production (Source: Robert Bosch GmbH).

Today, most pressure sensors are working using the piezoresistive principle combined with silicon micromachining. They are well suited for high volume production and system integration (Fig. 4). They have already found several automotive applications, such as the manifold absolute pressure MAP sensor, or are being in the process of series introduction. There is a general trend to replace existing pressure sensors based on other principles by piezoresistive MEMS sensors.

Especially, where high pressure is present, i.e. gasoline direct fuel injection, diesel common rail fuel injection, other resistive technologies (thin film or poly-silicon on steel) are used (Fig. 5) [5]. Thick film on ceramic diaphragm sensors are less common in automotive applications.

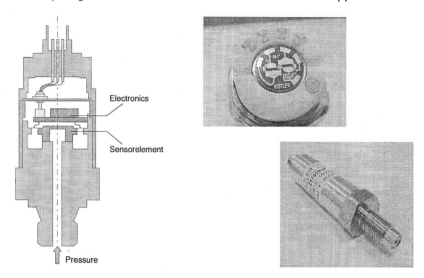

Fig. 5. Pressure sensor with sensing element fabricated using Thin Film or poly-Silicon on steel technology.

2.2 Capacitive Sensors

Capacitive pressure sensors consist of a parallel plate capacitor configuration and normally a vacuum gap between (Fig. 6). Typically, they use a thin diaphragm as one electrode plate of the capacitor. The other electrode plate is attached to a rigid substrate. An external pressure deflects the diaphragm, which in turn leads to a change of capacitance. A second reference capacitor without diaphragm may be used for compensation of thermal effects. Capacitive pressure sensors are realised in ceramic, or silicon technology. The sensor packaging is very similar to piezoresistive sensors. Capacitive sensors require dynamic excitation and contain an internal oscillator and signal demodulator to provide static output. They are best suited for measuring low pressures in the mbar range.

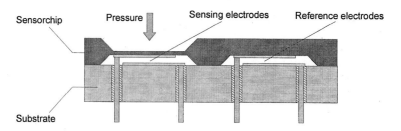

Fig. 6. Capacitive Pressure Sensor Element.

2.3 Piezoelectric Sensors

The basic piezoelectric measuring principle is shown schematically for a quartz pressure sensor in Fig. 7. The diaphragm converts the external pressure into a force which acts on the piezoelectric crystal. The stress inside the crystal generates an electrical charge on certain crystal surfaces, given by the crystal symmetry and cut. The charge is measured with charge amplifiers or high input impedance voltage converters. High sensitivity, high temperature range up to 400°C, high dynamic range, small size and high natural frequency are the principal advantages of piezoelectric measuring sensors. However, as active sensors they cannot provide true static measurements.

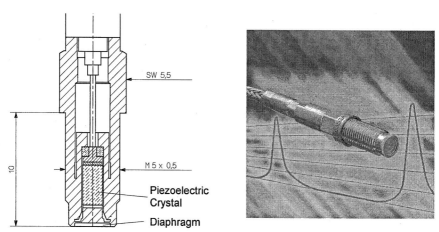

Fig. 7. Piezoelectric Cylinder Pressure Sensor.

Piezoelectric sensors are widely used for the measurement of quasistatic and dynamic pressure, especially in harsh environments. They represent the state of the art in combustion engine development and testing. Typical applications are high pressure measurements in the combustion chamber or in injection systems. Thanks to new high sensitivity crystals the pressure sensors can be miniaturised, in order to be integrated into a spark plug or glow plug. This represents a high potential for future automotive applications. However, applications in series vehicles are not found up to present due to the high cost of the measuring chain.

2.4 Resonant Sensors

Most resonant pressure sensors make use of a mechanical resonating structure (diaphragm, beam, cylinder, plate) which changes its natural frequency with applied force or strain. By using a piezoelectric crystal as resonating element, the excitation of the resonance (acoustic wave resonance) can be achieved directly by applying an oscillating voltage through the electrodes (inverse piezoelectric effect). Depending on the type of acoustic wave propagation, through the crystal (bulk) or on the surface, one speaks of bulk acoustic wave (BAW) or surface acoustic wave (SAW) devices. With micromachining technology such resonating structures may be directly fabricated on a silicon wafer [6], e.g. like resonating strain gages as shown in Fig. 8. Excitation and detection of resonance frequency can be achieved by electrostatic, piezoresistive, piezoelectric, inductive, magnetostrictive, magnetoresistive or optical principles.

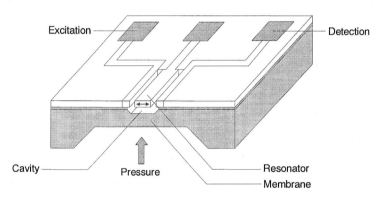

Fig. 8. Resonant PressureSsensor Element.

Resonant sensors have high potential for accurate measurements with high resolution and frequency output signal. SAW pressure sensors are passive, rugged and extremely small. They are well suited for wireless operation, e.g. pressure measurement in moving objects. For automotive applications resonant sensors, especially SAW pressure sensors, have high potential for wireless tire pressure measurement [7].

2.5 Optical Sensors

One common principle of optical pressure sensors uses modulation of light intensity by reflection at the diaphragm as function of the diaphragm displacement (Fig. 9, left). The optical fiber serves as light conductor, i.e. it transmits and receives the modulated light. Depending on the light signal processing, one, two or more fibers may be used. The light is usually generated by a light emitting diode (LED). Light intensity detection and conversion into a electrical signal is achieved with a photodetector (photodiode).

Another possibility to realize an optical pressure sensor consists of forming an interferometer (e.g. a Fabry-Perot interferometer) where the pressure causes a change of the interferometer length (either a cavity or an optical solid element) and modulates the phase difference between the reflected and interfering light waves (Fig. 9, right). Reflection occurs at the fiber end and at the diaphragm surface. For the interferometric principle the requirements for the light generation, fiber and light detection are much higher than for the intensity modulation principle.

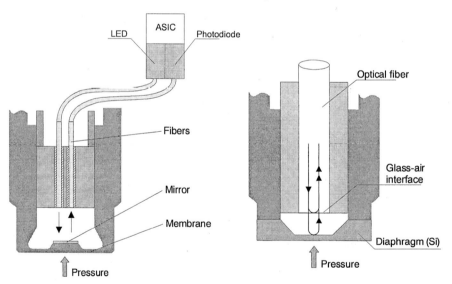

Fig. 9. Optical pressure sensor working with intensity modulation (left) and interferometric (right) principle.

There exist various other optical measuring principles. However, up to present they have only little importance for pressure sensors or are still in development.

Fiber optic pressure sensors have several inherent features which make them attractive for automotive applications: they are not susceptible to electromagnetic interference (EMI) and they can be miniaturized to very small dimensions below 2 mm diameter. However, optical fibers have to be handled with care and fiber connectors are critical components. In addition, optical systems have still higher costs and are therefore primarily found in research, development and testing applications. Automotive applications of optical pressure sensors are found for in-cylinder combustion pressure measurements [8, 9]. Due to their miniaturization potential they are well suited for integration into a spark-plug or glow-plug.

3 Conclusions

This overview has shown that depending on the specific requirements (technical specifications, environment, production volume, costs) of the application different measuring principles and sensor packages are used. Basically, it is distinguished between low volume research, development and testing applications and high volume series applications. In harsh environments like the combustion chamber pressure sensors are not yet introduced for series applications.
Clearly, MEMS pressure sensors are of growing importance. Especially piezoresistive sensors have proven to be mature for automotive series applications. They are replacing existing sensors based on other more traditional technologies or are finding new applications. In high pressure applications such as fuel injection or common rail systems robust thin film or polysilicon on steel pressure sensors are predominant. Piezoelectric or optical technologies are still in the low volume, high price market. They are potentially interesting for high pressure and high temperature applications in the engine environment (i.e. combustion or exhaust pressure). Here, recent developments of piezoresistive sensors with new materials (SOI, silicon carbide) are bearing high potential. Resonator pressure sensors are generally too complex in fabrication compared to piezoresistive sensors.
Future developments have to focus on reduction of over all system costs. The chosen sensor measuring principle and technology will depend on the environmental and reliability requirements given by the automotive application and on the possibility for high volume and low cost manufacturing.

References

[1] Peter Cockshott, Automotive Sensors, in Sensors, Vol. 8, Eds. W. Göpel, J. Hesse, J.N. Zemel, Micro- and Nanosensor Technology/Trends in Sensor Markets, Vol. Eds. H. Meixner and R. Jones, VCH Verlag, Weinheim, Germany, 1995, p. 491-523.

[2] E. Mounier, S. Leroy, M. Waga, Automotive Applications of MEMS: Overview of New Applications in Europe, Japan and North America and Evaluation of the Markets, in Advanced Microsystems for Automotive Applications 2000, Eds. S. Krüger, W. Gessner, Springer Verlag, Berlin, 2000, p. 1-16.

[3] P. Moulin, A. Akoachere, A. Truscott, A. Noble, R. Müller, M. Hart, G. Krötz, C. Cavalloni, M. Gnielka, J. Von Berg, Demonstrating the Benefits of a Cylinder Pressure Based Engine Management System on a Vehicle, in Advanced Microsystems for Automotive Applications 2002, Berlin (D), March 21-22, 2002, this volume.

[4] J. von Berg, M. Gnielka, C. Cavalloni, Th. Boltshauser, Th. Diepold, B. Mukhopadhyay, E. Obermeier: A piezoresistive low-pressure sensor fabricated using silicon-on-insulator (SOI) for harsh environment applications, Proc. TRANSDUCERS '01 Conference, Munich, June 10-14, 2001, p. 482-485.

[5] M. L. Dunbar, K. Sager, A Novel Media-Compatible Pressure Sensor for Automotive Applications, Sensors, January 2000, p. 28-31.

[6] T. Nishikawa, H. Kuwayama, S. Zager, High stable pressure sensor employing a resonant silicon sensor, ISA TECH/EXPO Technology Update Conference Proceedings (1997), Part 1/5 (of 5), Future Technology for Measurement and Control, Oct 7-9, 1997, Anaheim, CA, USA.

[7] A. Pohl, G. Ostermayer, L. Reindl, F. Seifert, Monitoring the tire pressure of cars using passive SAW sensors, Proc. IEEE Ultrasonics Symposium 1997, Vol. 1, 1997, p. 471-474.

[8] O. Ulrich, R. Wlodarczyk, M. Wlodarczyk, High-Accuracy Low-Cost Cylinder Pressure Sensor for Advanced Engine Controls, SAE paper no. 2001-01-0991.

[9] M. Fitzpatrick, R. Pechstedt, Y. Lu, A New Design of Optical In-Cylinder Pressure Sensor for Automotive Applications, SAE paper no. 2000-01-0539.

3.4.2 Tire Pressure Monitoring Systems – the New MEMS Based Safety Issue

J. Becker

Philips Semiconductors
Stresemannallee 101, 22529 Hamburg, Germany
Phone: +49/40/5613-3032, Fax: +49/40/5613-3045
E-mail: joerg.becker@philips.com, URL: www.semiconductors.philips.com

Keywords: Anti-blocking-system, pressure sensor, P2SC, TPMS, STARC, tire sensing

Abstract

After ABS and Airbag, a new safety feature will enter the majority of vehicles within the next five years - Tire Pressure Monitoring Systems (TPMS): MEMS devices, placed in the tires, measure pressure and temperature and send these information to a central dashboard unit in the vehicle by themselves, or by using other MEMS or microsystems.

The following paper describes the technical principal of so called direct Tire Pressure Monitoring Systems, using monitors embedded in the tire or embedded in a tire valve with included MEMS devices, micro-controller based signal conditioning chips and UHF transmitters to relay the relevant information to the vehicle. Also application specific challenges like automatic tire rotation detection are highlighted as well as the comparison with indirect (ABS sensor based) systems.

1 Introduction

Most car drivers ignore their tires, although they are one of the most crucial elements for the vehicle operation. Properly inflated tires aid the consumers in safety, performance, fuel economy and increased tire life. But one of five tires is under-inflated up to 40%. That leads to a significant tire life decrease and an increase of fuel consumption of around 1.5% by every 3PSI under-inflation. In spite of this, applications to monitor the tire pressure of the vehicle are in use only in niche markets for some luxury vehicles since a few years [1].

In Summer 2000, the company Firestone recalled more than 6.5 million tires, after the investigation, that some treads of their tires have separated caused by under-inflation. This prompted the US congress to enact the Transportation Recall Enhancement, Accountability, and Documentation (TREAD) Act on November 1, 2000. The TREAD Act mandates a rulemaking proceeding to require motor vehicles to be equipped with a Tire Pressure Monitoring System (TPMS) that warns the driver when a tire is significantly under-inflated.

The Act calls for the National Highway Traffic Safety Administration (NHTSA) to require Tire Pressure Monitoring Systems as standard on all cars, light trucks and multipurpose passenger vehicles sold in the USA from 2004 model year [2].

Outside of the US market, TPMS system are already established in high-end vehicles as optional feature and are becoming standard more and more.

Another driving force for TPMSs are so called runflat tires, which are so good at running at low inflation pressures, that a Tire Pressure Monitoring System is essential, to warn the driver in case of pressure loss.

2 Market Figures

Worldwide vehicle production will grow from 52 million passenger vehicles, SUVs and light trucks in 2001 to a projected 68 million vehicles by the year2008 - a Compound Average Growth Rate (CAGR) of 4%. The 2001 market for direct Tire Pressure Monitoring Systems was around 2.3 million Units. growing to expected 29.3 million Units in 2008. This is a CAGR of 44%. Indirect TPMSs will grow from 690 k Units in 2001 to 740 k Units in 2008 only [3].

Philips Semiconductors, the worldwide market leader in silicon for vehicle immobiliser and integrated remote keyless entry (RKE) systems is well positioned to serve this emerging market of direct TMPSs with silicon for current and future MEMS devices.

3.4.2 Tire Pressure Monitoring Systems – the New MEMS Based Safety Issue   245

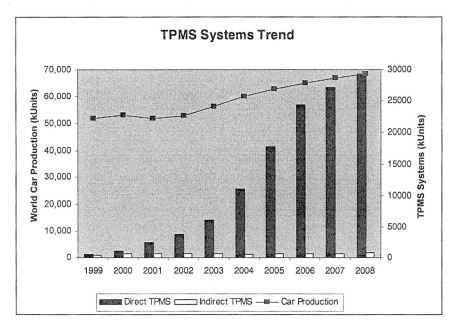

Fig. 1. TPMS Systems Trend

3 System Overview

A direct Tire Pressure Monitoring System is always measuring the pressure directly inside the tire. For this, Tire Modules are located in the tire, usually attached to the inflation valves. They broadcast their data via RF to a central receiver. This RF link uses the same RF principles and frequency range as the usually existing Remote Keyless Entry (RKE) system in the vehicle, thus it can be shared with the RKE system to reduce the system costs.

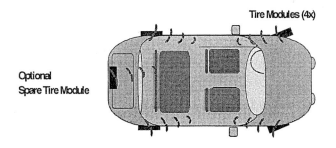

Fig. 2. Direct Tire Pressure Monitoring System

3.1 Tire Modules

The Tire Modules consist mainly of 3 parts:
- a (usually piezo-resistive analog) pressure sensor
- a so called Pressure Sensor Signal Conditioning Chip (which can be integrated in the sensor)
- a RF Transmitter Unit

The Tire Modules in the tire have to withstand temperatures from - 40 up to more than 150 degree Celsius in combination with acceleration ranges of 1000 g with crest to 2000 g. In this harsh environment, special components will be used to insure a lifetime up to ten years:

The Pressure Sensor

The pressure Sensor inside the Tire Module is a typical micro-electro mechanical system (MEMS) component. Currently, there are only few main specialist companies visible with the ability and experience in TPMS to build the sensors, robust enough to survive the tough conditions of real world driving. One of these is the company SensoNor from Horten, Norway [4], [5].

A currently used SensoNor pressure sensor is based on a so called "Triple Stack" element with a silicon diaphragm sandwiched between two layers of glass. The layers are vacuum sealed and customized to include metal conductors, electrical connections and any other components.

Pressure changes force the diaphragm to bow, producing a mechanical strain that other components can measure. The layering, vacuum sealing and a special transfer molding technique help protect the parts from moisture and corrosion.

The package, which also is a key element, may include the Pressure Sensor Signal Conditioning Chip, described in the next chapter.

Another important supplier for piezo-resistive analog tire pressure sensors is the company Novasensor from Fremont, California [6].

Signal Conditioning

Since the signal from the silicon sensor is of some µV, it has to be amplified and digitized. The full device has to be calibrated and initialized. For this, Philips Semiconductors has developed the P2SC, Pressure Sensor Signal Conditioning Chip.

It picks up the signal from the sensor bridge, changes it to digital, measures the temperature directly on chip and performs all the calibration

3.4.2 Tire Pressure Monitoring Systems – the New MEMS Based Safety Issue

and initialization needs. The P2SC includes the STARC based Reduced Instruction Set Computer (RISC) micro-controller core, which is field proven in Remote Keyless entry applications and also dedicated to TPMS.

Power consumption is optimized to the possible minimum, and as special feature the P2SC performs the wheel identification feature to solve the auto-rotation problem.

The RF Stage

Currently, external SAW or PLL based UHF transmitter, discrete or as an integrated device are used. But Philips Semiconductors already announced the second generation of the P2SC with the UHF PLL integrated on chip.

This will lead to a further reduction of cost and space on the PCB of the Pressure Sensing Module. Future devices will be fully integrated in the "Smart Sensor" housing - one chip - one package solution - in the tire.

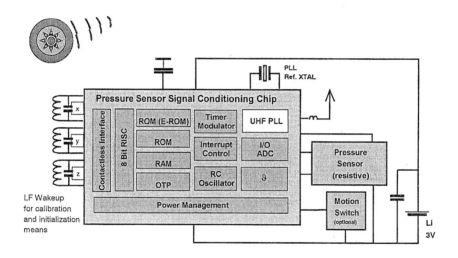

Fig. 3. Tire Pressure Monitoring Module

3.2 Receiver Module

As already mentioned, the receiver unit for TPMS is based on similar techniques like the already installed receivers for Remote Keyless Entry. So the existing RF receivers can be shared between TPMS and RKE. This leads to a significant cost reduction, thus a lot of car makers already forced their suppliers to integrate RKE and TPMS into one system. Philips Semiconductors and several other suppliers are offering RKE receiver ICs to address TMPS also [7].

4 Wheel Identification

After calibration and initialization, each tire is able to send its pressure information to the dashboard and the body controller knows, from which tire the signal is coming. But what happens, if the car driver is changing (rotating) the tires?

There a several ways to overcome this problem:
- Dedicated RF receiver for each wheel
- Inertial sensing of speed, combining with ABS/ESP information
- Amplitude analysis of the RF signals (RSSI)
- Bi-directional RF links
- LF wakeup

Philips Semiconductors has decided to use the LF wakeup for tire localization. This solution is fairly un-expensive and allows an immediate and reliable identification.

Small low frequency (125 kHz) wheelhouse antennas are sending a wakeup pattern to the specific tire module, which responds via the RF link. The LF wake-up has to bridge a distance from the wheelhouse antenna to the tire module of about 1 m, which is already proven to be feasible, using the Philips Semiconductors Passive Keyless Entry technology, where a similar distance has to be bridged to open a car remotely. A three dimensional (3D) interface in the Tire Module guarantees an orientation independent sensitivity for incoming wake-up pattern.

5 Indirect TPMS

Direct TPMSs are using tire pressure sensors. Indirect TPMSs rely on the presence of an anti-lock braking system (ABS) to detect and compare differences in the rotational speed of a vehicle's wheels. Wheel speed correlates to tire pressure since the diameter of a tire decreases slightly as tire pressure decreases.

If vehicles are already equipped with a 4 channel ABS system, the cost of modifying that system to serve the additional purpose of indirectly monitoring tire pressure would be significantly less than the cost of adding a direct TPMS to those vehicles. For vehicles not such equipped, adding a direct TPMS would be the less expensive way of monitoring tire pressure.

3.4.2 Tire Pressure Monitoring Systems – the New MEMS Based Safety Issue

In any case, direct TPMS are much more accurate than indirect systems, a comparison is given below [2]:

	Indirect TPMSs	Direct TPMSs
Cost of adding to vehicle with ABS, but without receiver	$12.90	$79
Cost of adding to vehicle with ABS and receiver	$12.90	$59
Cost of adding to vehicle without ABS or receiver	$130 for wheel speed sensors $240 for ABS	$79
Cost of adding to vehicle without ABS, but with receiver	$130 for wheel speed sensors $240 for ABS	$59
Susceptibility of wheel components to damage during tire installation and removal	Less likely	More likely
Need for an independent power source	No	Yes
Need to reset after a vehicle's tires are replaced or rotated	Yes, system must be re-calibrated	Yes
Ability to detect loss of air pressure if all four tires lose pressure	No	Yes
Ability to detect small pressure losses	No	Yes
Ability to detect under-inflated tire while vehicle is stationary	No, vehicle must be moving	Yes
Ability to identify which tire is under-inflated	No	Yes
Susceptible to giving false indications of a significantly under-inflated tire	Yes, if the vehicle is being driven on gravel or bumpy roads or at high speeds (70 mph) or if it has mismatched tires or a tire out of balance or out of alignment	No

Table 1. Comparison between Direct and Indirect TMPSs

6 Outlook

The current generation of Tire Pressure Monitoring Systems is based on a pressure sensor, which includes an ASIC for conditioning the pressure and temperature signal. The next generation, currently under development, will replace the ASIC by a micro-controller like the Philips P2SC.

Future Tire Pressure Monitoring Systems may make use of Bluetooth, when this technology will be established in the vehicle.
Since cost and lifetime of the batteries in the Tire Modules will remain an issue, the industry will keep on working on batteryless solutions, e.g. using inductive coupling our passive GHz technologies.

References

[1] http://www.schrader-bridgeport.net

[2] http://www.nhtsa.dot.gov

[3] Strategy Analytics, Automotive System Demand 1999 to 2008, July 2001 http://www.strategyanalytics.com

[4] http://www.sensonor.com

[5] Candace Stuart, Safety, Energy Efficiencies Pressure Tire Makers To Adopt MEMS Monitors, May3, 2001, Smartire Published Articles, http://www.smartire.com

[6] http://www.novasensor.com

[7] http://www.semiconductors.philips.com

3.4.3 Measurement of the Actuation Force of the Brake Pedal in Brake-by-wire Systems Via the Use of Micromechanical Sensors

E. Weiss[1], G. Braun[1], E. Schmidt[1], M. Dullinger[1], U. Witzel[2]

[1]BMW Group
Knorrstrasse 147, 80788 Munich, Germany
Phone: +49/89/382-30766, Fax: +49/89/382-46520
E-mail: elmar.weiss@bmw.de

[2]Ruhr University Bochum
Institute and Chair for Machine Elements and Design Theory
Universitaetsstrasse 150, 44780 Bochum, Germany

Keywords: actuation force, brake-by-wire, brake pedal, pressure sensor, standardisation, diaphragm

Abstract

This paper presents a new measurement concept for brake-pedal force in brake-by-wire systems by means of micromechanical pressure sensors. Differing from the conventional approach, the applied force is not measured indirectly via the pedal travel or angle, but directly at the point of application itself. First, the "driver – brake pedal" interface is analyzed so that the results can then be incorporated in a prototype brake pedal with an integral micromechanical sensor.

1 Introduction

The integration of new safety and comfort features in hydraulic brake systems is becoming increasingly more difficult and resource-intensive because of the fast information conversion that is required. This situation is an important driving force behind the rapid development of brake-by-wire-technology. [1]

At the man-machine interface (MMI) of today's vehicles, the force applied by the driver to the brake pedal is transmitted hydraulically to the brake components. This form of transmission is being superseded by the X-by-Wire-technology – electronic acquisition and further processing of the driver's intention is absolutely essential in this situation. The brake force can be detected by sensors using a variety of operating principles. The state of the art is indirect measurement of the driver's intention by evaluating the angle or path traveled by the pedal [2].

2   New Pedal Concept

To eliminate spurious effects and to maintain the redundancy that is absolutely essential for the safety components, the force required to actuate the pedal must be measured using a method based on another measurement principle in addition to the method of measurement using the pedal travel or angle. An obvious possibility is measuring the brake force directly at the point of application, namely the pedal surface. When this is done, the basic pedal geometry should not, and must not, be modified – the "feel" of the brakes is retained.

Apart from the redundancy mentioned above, this mode of measurement has a further important advantage. If the pedal is blocked by an object (e.g. a bottle or drinks can), it is nevertheless still possible to record the brake force because the measurement is not dependent on the pedal angle or travel. Vehicle braking can be carried out safely.

3   Analysis of the "Driver–Brake Pedal" MMI

Analysis of the man-machine interface is of special importance in order to select suitable sensors for installation on or in the surface of the brake pedal and to find the optimal method of attachment. Although some publications are already available on this topic [3], there are no known references relating to investigations into the force distribution over brake pedals. This key pedal parameter must, therefore, be investigated first of all:

Trials were, therefore, performed on production cars under normal operating conditions. A two-layered pressure measurement film from FUJIFILM (FUJI Measuring Film, Fig. 1) was used. The first layer (A-film) comprises a base layer with micro color-capsules; the second layer (C-film) is a base layer with a color developer. When pressure is applied to the film, the micro color-capsules are pressed onto the developer layer and color it red. The intensity of the coloration is proportional to the pressure applied to the film.

3.4.3 Measurement of the Actuation Force of the Brake Pedal

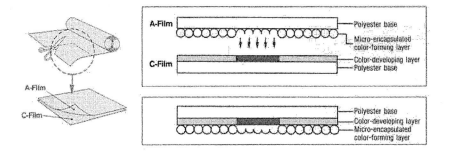

Fig. 1. FUJI Pressure Measurement Film [4]

A BMW from the latest 3 Series was used as the test vehicle. The original profile of the brake-pedal surface was covered with a filling compound and smoothed over. The brake rubber with the Fuji film inserted was then fitted to the brake pedal (Fig. 2). The purpose of the aluminium foil was to compensate for the unevenness on the inside of the brake rubber caused by the manufacturing process. The vehicle could then be driven as usual.

Fig. 2. Preparation of the FUJI Pressure Measurement Film for the Test
 a) Components (Brake Rubber, Thin Aluminium Foil, A-film, C-film)
 b) Assembled Pedal

Whenever there is a driver changeover, the test vehicle is fitted with a new pressure measurement foil. All the measurement foils that were used exhibited a pressure distribution that is similar to the one shown in Fig. 3: foot contact with the pedal is off-centre (always skewed to the right) because the driver must switch from the accelerator on the right to the brake on the left and chooses the most efficient movement. This can be confirmed by visual inspection of the replaceable brake rubbers on the pedals. The main area of abrasion is along the right-hand edge.

Fig. 3. FUJI Film after a Measurement Trip

4 Brake Pedal-prototype

As explained previously, the objective is to install the sensor for measuring brake force directly at the point of application of the force. In accordance with specifications, a micromechanical pressure sensor is to be used because the pressure is proportional to the acting force. On the basis of this requirement and the results from the MMI investigation, the pedal is now modified to permit sensor integration. At the same time, a search is made for sensors that meet the specifications.

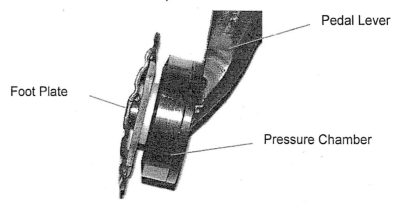

Fig. 4. Prototype Brake Pedal Incorporating Force Measurement

To integrate the pressure sensor into the line of action of the force, it should be installed flush with, or right under, the foot plate. The limited space available for installation is an important factor as the sensor system must not be allowed to impair the pedal function.

The requirements referred to above are addressed in a prototype pedal (Fig. 4). The force applied to the foot plate is converted into the pressure to

3.4.3 Measurement of the Actuation Force of the Brake Pedal   255

be measured. To transmit the pressure to the sensor still to be selected, the pressure chamber is filled with an incompressible fluid. Thanks to the internal design, it is immaterial whether the driver hits the center of the pedal or not – if the magnitude of the force is the same the pressure in the chamber is constant. This means that the requirements derived from Chapter 3 are met. The pedal lever itself is identical to the lever fitted to the brake pedals in today's vehicles so that the "feel" of the brakes remains the same.

5   Micromechanical Sensors

Micromechanical sensors are to be used in the pressure chamber of the new brake pedal because the installation space available is minimal. A few options for realization will now be discussed:

5.1   Sensors Constructed from a Novel Modular System

If the pedals are to be mass-produced cost-effectively at a later date, particular attention must be paid to the cost of the pressure sensor that is to be installed. In this context, an approach involving a modular system for pressure sensors that has already been presented in [5] could be adopted so allowing the elements that comprise the pressure sensor to be manufactured in large volumes. Monolithic integration of mechanics and electronics is also possible thanks to the design of the modular system.

The mechanical section of the modular system comprises the pressure diaphragm itself and various supplementary elements (Fig. 5) that make it possible to adapt to various applications and to drastically reduce the number of diaphragm variants. The most important supplementary element in the modular system is an element that allows the effective diaphragm diameter to be reduced. The diaphragm can be adapted to its intended use (e.g. greater pressures) by means of an underlay element with a simple drill-hole. It behaves like a diaphragm with a minimal diameter, the pressure vs. bending curve remaining linear. Further supplementary elements in the modular system should comprise underlay elements which, as well as having a drill hole, should also be chamfered around the drill hole. The chamfering can be linear or curved. Two or three level chamfering could also be considered.

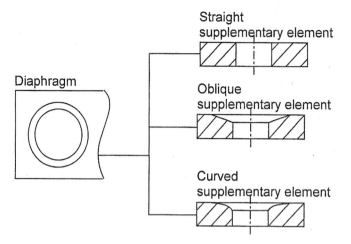

Fig. 5. Micromechanical Pressure-Diaphragm Modular System

The sensor to be integrated into the pedal should meet two requirements: in the lower, "normal" measurement range (to full braking) the measurement should be as accurate as possible; in the upper "excess pressure range" (to pedal failure) no further results are required because the vehicle is already being decelerated with the application of maximum braking. However, the sensor must not be damaged in this situation as it must still be capable of functioning properly when the pressure is removed. On the basis of these technical requirements, a pressure sensor should be constructed from the modular system referred to above.

A standard diaphragm is the basic element in the modular system. The diaphragm bending and so the measurement curve is linear over the whole measurement range (Fig. 6, Diaphragm without underlay element). The required measurement curve is, however, not constant over its whole range, but flat in the second section (where no further measurements are required) - in other words, the standard diaphragm must be adapted using suitable supplementary elements so that it can perform its intended function.

Modular system calculations show that an element with a chamfer has a curve that flattens off and so has similarities with the required curve. The curve is shown in Fig. 6, "Chamfered underlay disk" alongside the ideal curve. As required, the first section of the curve is almost linear, which means that measurements of sufficient accuracy can be made. The curve then flattens out considerably (accurate measurements are no longer required), but still continues to increase. The second section of the curve

3.4.3 Measurement of the Actuation Force of the Brake Pedal

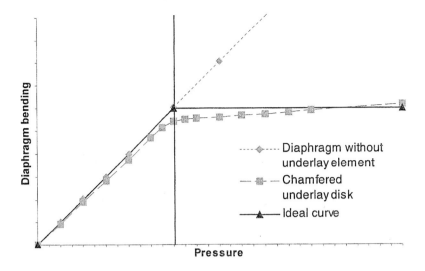

Fig. 6. Pressure vs. Bending Curve of the Micromechanical Diaphragms from the Diaphragm Modular System

however meets specifications as the strength of the diaphragm is still guaranteed in spite of the overloading.

The modular system that has been described is, therefore, a simple way of constructing a sensor from standard components. Thanks to modular design, adaptation to even complex specifications is possible. At present, intensive efforts are being made to produce a prototype of the brake-pedal sensor described above.

5.2 Standard Sensors

In order to perform function tests on the pedal prototype immediately, mass-produced sensors from a variety of manufacturers are now being fitted to the pedal. These pressure sensors must meet the requirements specified in Chapter 4.1. Standard micromechanical sensors operating on the following measurement principles are used:

Strain Measurement Strips (SMS)

The thin-film SMS sensor (Fig. 7) on metal diaphragms is a very compact but expensive sensor (diameter = 6 mm, h = 0.6 mm). Positive features are its insensitivity to vibration and mechanical shocks and a long operating life. This sensor is also insensitive to pressure media that are not too aggressive. A drawback of the SMS principle is the minimal strain sensitivity (10 times less than that provided by the piezo-resistive approach).

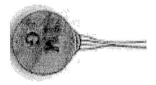

Fig. 7. SMS Sensor

Fig. 8 below shows the voltage vs. force curve from SMS sensor measurements on the brake test rig. Forces were applied to the pedal in steps and the associated output signals determined. It is obvious that the same output signal is obtained when an identical force is applied. The measurements are repeatable – the curves almost coincide. However, the curve is not linear. As the sensor output signal is very weak, an amplifier is required.

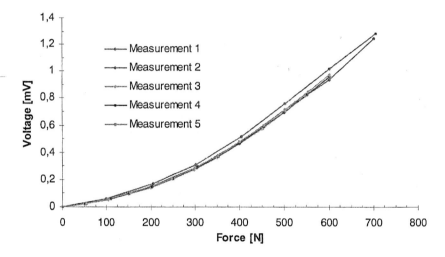

Fig. 8. Results produced by an SMS Sensor

5.3 Piezo-resistive Sensors

Piezo-resistive pressure sensors are currently the most popular solution to industrial measurement and control problems because the development of semiconductor technology and silicon-based micromechanics has made it possible to manufacture piezo-resistive pressure measurement cells in large volumes at decreasing costs. High accuracy and fast response times are two advantages of this technology.

3.4.3 Measurement of the Actuation Force of the Brake Pedal

Fig. 9 shows the results of measurements on a piezo-resistive pressure sensor incorporated in a pedal. The measurement procedure is the same as that used for the SMS sensor. In this trial too, the measurements had a high level of repeatability. The measurement curves are highly linear. This means that when vehicle trials are performed, it will be easy to assign measured pressure and applied force without much additional effort.

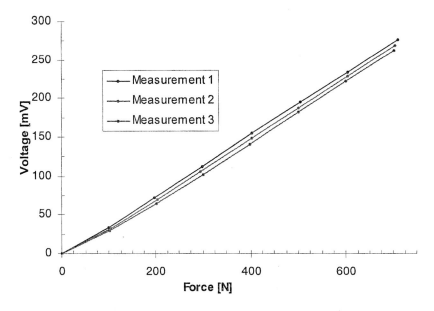

Fig. 9. Results Obtained with a Piezo-resistive Sensor

6 Summary and Prospects

The direct measurement of pedal force at the point of application has a number of advantages. For this reason, this paper presents a prototype brake pedal that measures pedal force using micromechanical pressure sensors. For mass production, it is absolutely essential to use cost-effective sensors. In this context, sensors constructed from a modular system whose components can be manufactured in large volumes at attractive prices could be employed. A sensor with the required characteristics can then simply be put together cost-effectively from the various elements that are available. An implementation of the modular system principle and subsequent tests are being prepared. However, in order to check the characteristics of the pedals now, the first step is to integrate standard sensors based on a variety of physical principles into the pedal and to perform trials. No matter which sensor was selected, the brake test jig showed that the pedal/sensor combinations functioned well; the repeatability of the results was high and further tests can be carried out in vehicles.

Acknowledgements

The authors would like to thank "Graduiertenkolleg 384: Nanoelectronic, micromechanical and micro-optical systems" at the Ruhr-University, Bochum and its head Prof. Dr. A. Wieck for their support. Thanks also go to all BMW employees who have provided help and advice to the authors in relation to the work that has been completed so far. Jakob Unterforsthuber, Gerhard Vielwerth, Werner Stehbeck, Stefan Klein and Marc Kühn can be singled out for special thanks for their assistance.

References

[1] N. N.: "Neues Konzept für Bremsen", Blick durch die Wirtschaft, issue dated 11.09.1997.

[2] H. T. Dorißen, D. Hobein, H. Irle, N. Kost: "Beruehrungslose Weg- und Winkel-Sensorik – Innovative Loesungen für automobiltechnische Anwendungen", ATZ – Automobiltechnische Zeitschrift, volume 10, 1998.

[3] N. N.: „Forschungswerkzeug zur Untersuchung der Schnittstelle Fahrer / Bremspedal", ATZ – Automobiltechnische Zeitschrift, 01.02.1999.

[4] N. N.: "FUJI Measurement Film", instructions accompanying the FUJI pressure measurement film.

[5] E. Weiss, E. G. Welp, U. Witzel, A. Wieck, E. Schmidt: "Investigation into the standardization of micromechanical components and their simulation and computation using FEM - case study of diaphragms", MSM2001, Hilton Head Island, South Carolina, USA, MSM, Vol. 4 (2001).

3.4.4 Design of a New Concept Pressure Sensor for X-by-wire Automotive Applications

L. Tomasi[1], G. Kroetz[1], E. Wieser[1], W. Suedkamp[2], P. Thiele[2], E. Obermeier[3]

[1]European Aeronautic Defence and Space Company Deutschland GmbH
Corporate Research Center
Microsystems Department, 81663 Munich, Germany
Phone: +49/89/607-24029, Fax. +49/89/607-24001
E-mail: luca.tomasi@eads.net

[2]Aktiv Sensor GmbH
Ruhlsdorfer Strasse 95, Gebaeude 4, 14532 Stahnsdorf, Germany
Phone: +49/3329/606831, Fax. +49/3329/606815
E-mail: wsuedkamp@aktiv-sensor.de

[3]Technische Universitaet Berlin
Gustav-Meyer-Allee 25, 13355 Berlin, Germany
Phone: +49/30/314-72769, Fax. +49/30/314-72603
E-mail: oberm@mat.ee.tu-berlin.de

Keywords: pressure sensor, self-test, two levels sensitivity

Abstract

Although the automotive market is slowly moving towards electro-mechanical X-by-Wire systems for low and middle class vehicles, there is still a significant need for electro-hydraulic systems in high power braking and steering applications. Therefore, an ongoing development in the direction of more precise, reliable and cost effective systems is taking place. Consequently, the performances of the respective sensorial systems have to enhance. Hence justifying the development effort for new concepts for MEMS pressure sensors. Moreover, many of the servo-assisted electro-hydraulic applications operate in two pressure ranges: a low range, where high precision is needed, and a high pressure one, where stability is a primary requirement. Normally these characteristics can only be fulfilled by using two pressure sensors with two different ranges. In this paper, a silicon pressure sensor with a two levels sensitivity that aims to meet the described specifications in one sensor design is presented. Moreover, self-test and recalibration are conceived in the design in order to enhance reliability and precision at a cost effective way.

1 Introduction

Growing wealth, rising living standards and human mobility increases the demand for comfort and safety in transportation systems. Though the recent statistics report a decreasing number of fatal accidents in road traffic, the percentage of casualties with passenger cars is proportionally growing: 60% was the quota for Germany in 1997 [1]. At the same time the effort of introducing passive safety in automotive since the sixties (invention of the safety belt) has been so intense that it will most likely reach the saturation of its possibilities very soon. To further increase safety there is therefore the need for accident avoidance strategies (active safety), of which ABS (Antilock Braking System) and EPS (Electronic Stability Program) have been the first market available examples. In the near future a progressive introduction of active safety systems in cars is to be foreseen, leading, on the long term, towards an increasing driving assistance [2].

An unavoidable step in the direction of active safety is mechatronic, an integration of sensors and electronically controlled actuators: X-by-Wire (managing a system through electrical wires) is part of this. X-by-Wire in automotive allows driver-independent applications that enhance performances, stability and system diagnosis, improving safety and comfort. Particularly interesting applications of X-by-Wire in safety relevant fields are steering and braking systems. Concepts for electro-mechanic braking and steering are already being developed [3], [4]. The Introduction of these systems, though, is not to be expected on the short term as the present vehicle electric system (the battery and the other related parts) is not able to provide the needed power: this is particularly true for braking systems. On the other hand an electro-hydraulic braking system combines flexibility and diagnosis of mechatronic with the hydraulic backup for braking power accessibility. The first reported market available product is the Sensotronic Brake Control (SBC) in the Mercedes SL class in 2001: result of the development cooperation effort of Daimler Chrysler AG and Robert Bosch GmbH [5]. In order to ensure safety, such systems have high precision and availability (reliability) requirements. These specifications reflect on the system subparts also, sensors included. Hence, there is a need for new pressure sensors with an increased precision for a high volume market and thereby justifying the development effort described in this article: the SBC, for example, makes use of six pressure sensors, multiplied by the automotive volume production this gives good idea of the market needs. Moreover, five of these sensors work in a high-pressure range (up to 200-250 bar) and one in a lower range as it is placed on the braking pedal, where the so-called "driver braking whish" is determined, therefore a high precision is needed in this range since the pressure building up in the early phases of braking is rather slow. At the same time the pressure measurement has to be correlated for redundancy to the pedal position sensor output, which on the other hand increases quite rapidly as the driver starts braking. Higher precision enables also faster system diagnosis. In this paper is presented a pressure sensor, which combines all the required characteristics in one transducer. This sensor is conceived to combine an intelligent sensor design with the advantages of digital electronics, following concepts in part previously presented in some other AMAA contributions [6], although avoiding the limitation of all integrated solutions for high pressure applications and never the less going in the direction of what Giachino already in 1986 defined as "smart" sensors: integrated or not [7].

2 Concept for the Pressure Sensor

The major requirements of a pressure sensor for X-by-Wire applications, as previously mentioned, are high precision and reliability as well as multi-functionality and flexibility (features strongly desired in modern sensor design). These requirements have heavily influenced the design choices. In order to enhance the precision, it has been conceived a silicon micro machined piezo-resistive pressure sensor chip with two different sensitivities: a higher one in a low-pressure range (0 to 30 bar), where often an elevated resolution is required, and a lower one at higher pressures (up to 250 bar). Thus, with one single membrane chip, practically two sensors are obtained. Moreover, as it will be explained further on more in details, the transition between the two sensitivity levels determines an area with particularly interesting characteristics that could be used to recalibrate the sensor from offsets without having to remove it from the system where it normally operates and mount it on a reference bench. Somehow what could be called a "self-recalibration" ability.

Enhancing the reliability and therefore the availability of a sensor requires stability in the components and sensor health monitoring strategies. This latter is possible through an integrated digital electronic that would allow self-test functions. Key point of these procedures is the previously mentioned recalibration area, which potentially allows monitoring offsets with a precision up to 0.15% full scale (FS) without need of integrated actuators and relative control electronics. Further digital electronic can be designed, without major difficulties, to integrate a controller for networking, consequently enhancing capabilities and flexibility of the sensor.

2.1 Two Levels Sensitivity and Recalibration

The transduction of the physical quantity, pressure in the specific case, into an electrically measurable figure is performed through piezo-resistive elements implanted into the surface of the silicon chip. These transducers are sensitive to the stresses in the two coordinates defined with respect to the plane where the elements are implanted in the chip [8]. Stresses on the piezo-resistors induce changes in their resistance that can be detected with rather high accuracy as unbalance of a Wheatstone bridge. The stresses on the chip surface depend on the geometrical characteristics of the latter and on the forces deriving from the applied pressure [9]. Therefore transducers are usually placed in such a way to have maximum response to the pressure changes and to obtain a constant sensitivity. Normally small variations in the sensitivity are undesirable as they complicate the calibration process and often reduce the sensor accuracy. On the contrary, in the presented design, an abrupt change in the sensitivity has been conceived through a major variation of the sensor geometry. This characteristic has been exploited to realize the two sensitivity ranges.

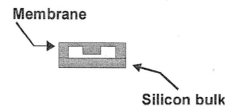

Fig. 1. Diagram of the Sensor Chip Profile

The sensor consists of a membrane at which centre is placed a cylindrical structure, as shown in Fig. 1. As the pressure is applied, from top, the membrane will move freely downward: this determines a rather sensitive sensor response, which will continue until 30 bar is reached. At this point the cylinder will enter into contact with the silicon bulk plate. Consequently, the geometrical structure of the sensor will almost instantly change: the membrane will not be able to move freely any more and will behave rather like a ring fixed at the two sides. The stiffness of the structure will significantly increase, thus the building up of stresses due to pressure will reduce and thereby the sensitivity will be roughly one quarter than the one between 0 and 30 bar. This determines the low sensitivity range that is specified up to 250 bar (Fig. 2). Moreover the cylindrical central structure makes the membrane fairly robust and resistant to overpressures.

In silicon the elastic behaviour, opposed to the plastic one, is dominant. Therefore, silicon withstands stresses with almost unchanged characteristics: this is what makes it a good material for mechanical sensors. Thus, it can be expected that in the described design the cylindrical central structure and the respective contact area on the silicon

3.4.4 Design of a New Concept Pressure Sensor

bulk will remain stable. Consequently it can also be expected that the pressure needed to generate the contact between the two parts will remain constant through the sensor lifetime, thereby the transition between the two sensitivity levels will take place always at the same pressure: in Fig. 2 this is defined as Recalibration point.

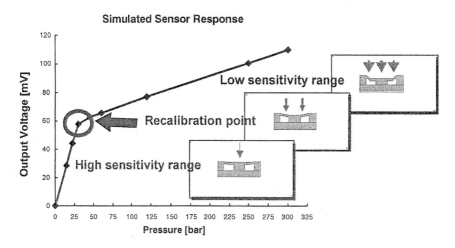

Fig. 2. Sensor Response Extrapolated from FE Simulation. The three mechanical configurations are also shown.

Now, gathering this information together, a contact point is obtained, which is mechanically determined, constant and independent from the electrical characteristics of the transducers. Therefore, if it is possible to evaluate a procedure to determine this point through the normal sensor operation, than a monitoring and correction of electrical instabilities such as offset drifts can be achieved without need of a reference sensor or external action: a simple example of how this could be obtained will be given in the paragraph 2.2. Moreover, this recalibration principle does not need any internal actuation system: no actuator control or extra technology is therefore needed. The sensor integrates what can be called a passive recalibration and self-test principle.

Least but not last, the contact or recalibration point is determined through the sensor technology and can be therefore defined differently from sensor to sensor. In the case the sensor is operating in a network environment where other sensors with different contact pressures are present, it is possible to obtain more recalibration points, potentially increasing the sensor accuracy.

2.2 Integrated Digital Electronic and Self-test

Digital electronic is often thought to be expensive for pressure sensors. This argument usually does not consider all the potential advantages that it can bring, either because of the difficulty to have a complete overview on them or as a rather significant research effort is needed to be able to exploit them completely. Moreover costs of digital electronic are on the long term continuously decreasing.

In the presented design digital electronic was introduced to implement monitoring and correction strategies in the sensor. Already at design level activities were carried out to investigate all possible failures of the sensor and evaluate their entity, with the goal of eliminating as much of them as possible through design, particularly those that could not be automatically detected by the sensor. On the remainder will be, in the first place, evaluated methods to individuate the errors (self-test) and, when possible, correct them without outside intervention (recalibration). A diagram of this procedure is described in Fig. 3.

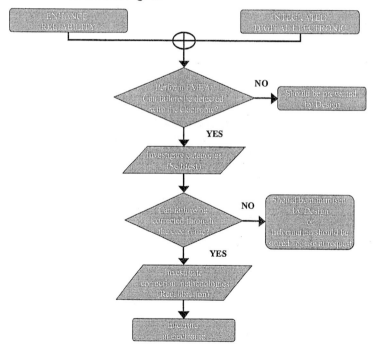

Fig. 3. Self-test and Recalibration Flow Diagram with Digital Electronic

Furthermore network capabilities can be introduced and thereby user tailored functions can be programmed resulting in an enhanced sensor flexibility.

3.4.4 Design of a New Concept Pressure Sensor

Clearly, a complex electronic has not only advantages; therefore consideration has to be taken not to introduce further hardware or software errors.

Central point of the self-test strategies is the previously described "Recalibration point". The presence of digital electronic allows performing internal drift monitoring and recalibration. A simple example might help the understanding.

Lets suppose that the sensor is working in a system where the pressure can rise linearly, namely 250 bar in 8 sec., for simplicity lets also suppose that the sensor has an ideal linear behaviour in the two sensitivity ranges (in the real case there will be a linearity error which will ad up to the calculations, on the other hand though, the sensor response could be better described by polynomials of higher order, therefore it has been chosen to stay with the simplest case). During the pressure rise four points are sampled through the digital electronic: point one at sensor output around 0 V and the second around 2 V, in the low pressure range, the third at 2.3 V and the fourth at 4 V, in the high pressure one as shown in Fig. 4 (a wise choice of the points can influence up to 50% the accuracy with which the recalibration point can be determined). These points are used to define the two lines, which intersection will determine the contact voltage. This can be compared with the value stored in the sensor memory at the previous recalibration and, if the difference exceeds the calculation errors, the new value will substitute the old one: the sensor response lines will be adjusted and thereby a recalibration will take place. Key point of this procedure is the dimension of the calculation errors. If the linearity error is not considered, for the reasons previously given, these inaccuracies depend on the sensor A/D converter resolution and the sampling frequency. Therefore, with a 10 bit A/D converter and sampling at 1 kHz a recalibration with approximately a 0.15 % accuracy FS can be obtained.

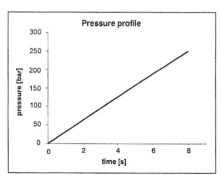
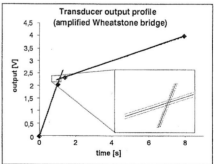

Fig. 4. Example of Possible Calibration Procedure

3 Sensor Design

Defining a concept for a new sensor is no trivial job. Putting this into a realisable design is even more complex and requires a good deal of experience in sensor manufacturing and simulation techniques. The transducer chip design has been conceived in collaboration between EADS (European Aeronautic Defence and Space Company) Deutschland GmbH and AKTIV SENSOR GmbH, with the contribution of the Technical University of Berlin. The electronic design instead was the result of the cooperation of EADS Deutschland GmbH and ELBAU GmbH.

3.1 Chip Design

The major difficulty in the design was to realise the change in the mechanical structure in such a way that the sensor response variation between the two configurations would be possibly sharp, but most of all that the response with respect to the pressure change would be monotonous. If this condition is not fulfilled, there is no one to one correspondence between the transducer response and the applied pressure: there will be different pressures that will produce the same output signal, thereby the sensor will be intrinsically unreliable and therefore unusable. Overcoming this problem means that the piezo-resistors (the transducing elements) have to be exposed to always increasing stresses with the rising of the pressure. Therefore, the choice on the piezo-resistor position on the chip membrane is determinant and with it the results of the simulation. In Fig. 5 the choice that has been made in the positioning of the piezo-resistive elements is shown: it can be noted that the stress distribution changes significantly before and after the mechanical contact.

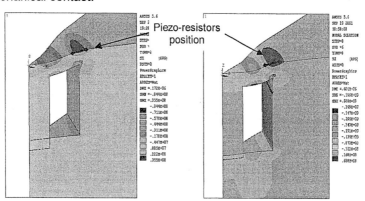

Fig 5. Simulation of the Stress Distribution on the x-axis on the Sensor Chip at 23 Bar, Before the Mechanical Contact, and at 300 Bar, the Maximum Pressure

Moreover, it has been chosen to design 90-degree profiles in order to reduce the previously described risk, implying the use of anisotropy etching. The results of the dry etching process can be seen in Fig. 6.

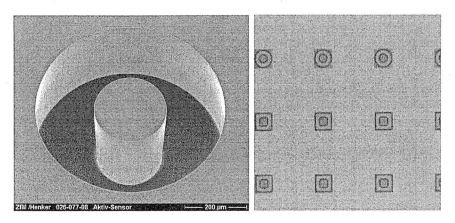

Fig. 6. SEM Picture of the Chip Structure and X-ray Picture of the Bonded Wafer Showing the Circular and Square Sensor Design

3.2 Electronic Design

Design of the electronic should be maintained to a low level of complexity. Never the less attention should be given in designing to enable the implementation of all self-test and recalibration features allowed, but at the same time avoiding unnecessary over dimensioning of components that would only reflect itself on an increase of costs. Particular care should be given in taking advantage of the high resolution in the low-pressure range: for example, in the case of a linear analogue or Pulse-Width Modulation (PWM) output is desired, as it normally is in sensor output coding, a high resolution digital to analogue converter is needed. Moreover, in the design is planned: a volatile memory for storing the calibration parameters, a non-volatile one for the programming of the self-test and recalibration algorithms, a PWM module, a CAN (Controller Area Network) module for bus communication and of course analogue to digital converter to enable the signal processing.

In the first prototype a low level of integration has been chosen to enable more design flexibility, never the less most of the needed functions could be performed by a commercially available ASIC which could be integrated in a second stage.

4 Conclusions

In this paper has been presented a novel concept for pressure sensor design that goes well beyond the sensor integration problematic, proposing new ways to exploit silicon bulk micro machining for enhancing precision and reliability introducing a multi-resolution concept and self-test features opening doors to applications not only in the automotive field.

The first prototype has been realised and the first results are encouraging and maybe will be argument of a future AMAA contribution.

Acknowledgements

The author would like to thank all the partners of the MuSe project that, with their intense work, gave birth to the material for this article and all those that, with their comments and suggestions, have improved the quality of this paper.

References

[1] K. Rompe, "Die Notwendigkeit aktiver Sicherheitssysteme im Fahrzeug", EUROFORUM Konferenz Aktive Systeme zur Fahrzeugsicherheit, Köln, 3.3.1999.

[2] P. Rieth, "Technologie im Wandel X-by-Wire", IIR Konferenz Neue Elektronikkonzepte in der Automobilindustrie, Stuttgart, 13-14.4.1999.

[3] H. A. Beller, et al., "Total Chassis Management – Heading for the Intelligent Chassis", Proceeding of the XXth brake conference, VDI, October 2000, pp. 150-178.

[4] B. Bayer, "Brake-by-wire – Ein elektronisch-mechanisches Bremssystem", IIR Conference X-by-Wire, Frankfurt am Main, 16-17.10.2001.

[5] O. Booz, et al., "Electro-Hydraulic Brake", IRR Conference X-by-Wire, Frankfurt am Main, 16-17.10.2001.

[6] R. Schmidt, et al., "Pressure Sensors for Automotive Applications – with New One chip EPROM Technology", AMAA 2001, pp. 131-138.

[7] J. M. Giachino, "Smart Sensors", Sensors and Actuators, Volume 10, 1986, pp. 239-248.

[8] Y. Kanda, "A Graphical Representation of the Piezoresistance Coefficients in Silicon", IEEE Trans. Electron Devices, Edition 29, 1982, pp. 64-70.

[9] S. P. Timoshenko, et al., "Theory of Plates and Shells", Mc Graw-Hill, 1987.

3.4.5 Pressure Sensors in the Pressure Range 0-300 bar for Oil Pressure Applications – Directly in the Oil Media

R. Schmidt

Fuji Electric GmbH
Lyoner Strasse 26, 60528 Frankfurt, Germany
Phone: +49/69/669029-22, Fax: +49/69/669029-56
E-mail: rschmidt@fujielectric.de, URL: www.fujielectric.de

Keywords: calibration after assembled, digital trimming, double diaphragm package

Abstract

This paper will focus on pressure sensors which measures the pressure directly in the oil media, like CVT (continuos variable transmissions) applications, by using piezo resistive sensor chips with 48 bit EPROM trimming technology, all in one chip solution, integrated EMI filter and a special double diaphragm package. So the sensor can remain directly in the oil area.

1 Introduction

Fuji Electric is developing a pressure sensor based on the new sensor chip – EPROM technology to measure the oil pressure directly in the oil media. The required output characteristic is realised by a digital trimming / programming 48 Bit EPROM. The features of the EPROM technology are: higher accuracy solution by calibration after assembling a simple structure by all in one chip design, (includes EMC filter – 100 V/m and surge protection), easier production process and suitability for wide application (low to high pressure).

1.1 One Chip Design

The chip size is 3,2 mm x 3,2 mm. A p-type silicon wafer (6" wafer size – CMOS IC process) is used for a substrate. A diaphragm is fabricated by chemical etching. On the surface of the diaphragm, four strain gauges (in a wheatstone bridge) are diffused in the p-type direction. Although the signal strength is in the few tens of millivolts and there is a strong dependence due to the amplification circuit and temperature compensatory circuit, Fuji Electric is able to maintain a linear relationship between the output voltage and the mechanical strain, allowing an accurate control of the circuit. The sensor can cover wide operating pressure range by a variation – optimising the thickness and diameter of the diaphragm, which can be changed by altering the etching time (up to 700 bar). Integrated functions are a sensing gauge / diaphragm connected in a wheatstone bridge, an analogue signal amplifier circuit for DC output voltage, digital signal adjustment circuit of wide temperature range, surge protection circuit for ESD and over voltage, diagnostic circuit and EMI filter circuit etc. To achieve higher accuracy, the new digital calibration method using EPROM and an optimised process has been developed.

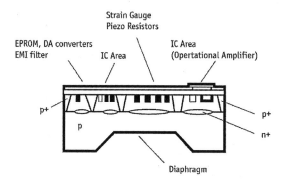

Fig. 1. Sectional View of the Chip

The four piezo resistors connected in a wheatstone bridge to detect measuring pressure, are positioned at the centre area of the chip and the other circuits elements like analogue signal amplifier, digital signal conditioning and EMC filter circuit etc are located around the diaphragm.

1.2 Circuit Design and Sensor Structure

The integrated temperature sensor is used for the compensation of temperature characteristic. The digital circuit of an EPROM, shift register and DA converters is newly designed for a more simple, easier and higher accurate calibration method. The adjustments are carried out for 5 parameters which are span (sensitivity), offset, span temperature characteristics, offset temperature characteristics and offset temperature characteristics of the temperature sensor.

After assembling, every sensor cell is measured by two temperature characteristics and calibrations to the target characteristics of 0,5 up to 4,5 V output voltage at 5 V power supply condition by changing an external signal. Then a final test with three – pressure point and three - temperature point measurement is carried out. A digital trimming method using EPROM provides a high accuracy solution because the digital value of EPROM does not change during the accelerated high temperature storage test of 250 degree for 500 hours.

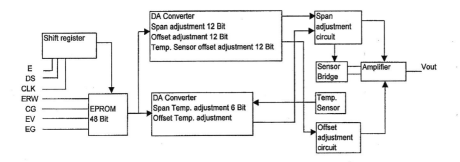

Fig. 2. Block Diagram and Trimming Method

2   Pressure Measuring Directly in the Oil Media

The pressure range is 0 – 10 bar, 0 – 30 bar, 0 – 50 bar and in a second step up to 300 bar. Typical applications are: pressure sensors which measures the pressure directly in the oil media like CVT / Continuous Variable Transmissions. The sensor is used to measure hydraulic controlled pressure in automatic transmissions for passenger cars. The sensor converts oil pressure in an electrical signal, which provides the information to the transmission electronic controller. This sensor is

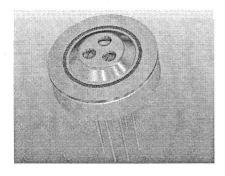   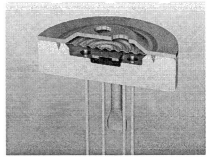

Fig. 3. Oil Pressure Sensor   **Fig. 4.** Sectional View

integrated in a base plate of an oil tight electronic controller and remains in the oil area. It can be fully immersed in oil or in an oil environment.

To realise the media selection, Fuji is using a double diaphragm package. The oil gets through 3 holes onto the second diaphragm / package. This area is stuffed with silicon oil. The pressure is transmitted by this silicon oil to the EPROM semiconductor chip. So this sensor can be mounted directly in the oil media and remains in the oil area.

2.1   Design for Noise Prevention / EMI Filter

Conventional noise prevention measures required circuit components, such as capacitors, feedthrough capacitors, inductors, a circuit board for electrical connections, and a shield case. With the EMI-prevention pressure sensor, these externally mounted circuit components and assembly processes for noise prevention are not necessary. The low pass filter for noise reduction is constructed with capacitors and thin film resistors. However, a MOS process is utilised for formation of the capacitors by using a high quality oxidized film. A Capacitor – Resistor filter is connected between the power supply voltage input pin (VCC) and the GND pin, and between the signal voltage pin (VOUT) and the GND pin. This filter reduces the input noise which came in from the wire harness.

A MOS capacitor, which was used as a capacitor for low pass filter, is close to an ideal capacitor and their frequency characteristic is very good. Fig. 5 shows those frequency characteristics. A thin film resistor is used for the low pass filter. This resistor has a low parasitic capacitance and good high frequency characteristics.

Frequency characteristic of the low pass filter and the cut-off frequency are shown in Fig. 5 and Fig. 6. Above the frequency of 10 MHz, the input noise caused of electromagnetic waves has a direct influence on the wire length of the filter. Therefore the cut-off frequency of the filter has to be designed to 10 MHz. This filter (second degree) was selected so that the attenuation characteristics of the low pass filter would have a steep slope.

To suppress voltage drop due to the filter at the power supply voltage input pin, the supply currently is suppressed with a stain gauge unit, operational amplifier, compensation and adjustment resistors. As a result, an output operating rating of 0,2 to 4,75 V can be maintained by a power supply of 5 V. To protect the capacitor of the low pass filter, a zener diode is connected to input and output pins to absorb static electricity and surge voltage. This guarantees the necessary durability for automotive applications.

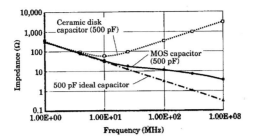

Fig. 5. Frequency Characteristics of MOS Capacitor

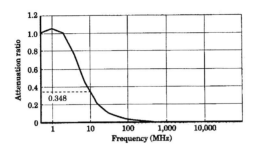

Fig. 6. Cut-off Frequency of a Low Pass Filter

3 Conclusion

Fuji Electric's main target is to develop and produce low cost pressure sensors by using advanced 1 micron full 6' CMOS design and all in one chip solutions with digital adjustment by EPROM. A further advantage of this new sensor is a higher accuracy and a simple structure. Same chip technology can be used for some other automotive applications like MAP sensors, Brake Systems, Gasoline Direct Injection, Exhaust Gas Recirculation etc.

3.5 Further Functions and Applications

3.5.1 Sensors in the Next Generation Automotive Networks

G. H. Teepe

Motorola GmbH
Schatzbogen 7, 81829 Munich, Germany
E-mail: gerd.teepe@motorola.com

Keywords: networking, bus topology, sensors, multiplexing, LIN, CAN, FlexRay

Abstract

This paper will focus on the future integration of sensor systems into the car electronics architecture. Today sensors are mainly equipped with an electric ratiometric output in voltage current or frequency. The signal conditioning is usually performed as part of the electronic control system and specifically tailored to suit the needs of the sensor.

As automotive electronics migrates towards decentralized, bus based architectures, the urgent need for the sensors to fit seamlessly into this new system architecture arises. This paper contributes to the discussion on how to best move forward with sensor interfaces in view to create a robust sensor infrastructure in the car.

1   Introduction

The choice of a bus-structure in the car is a serious issue for car makers with very long-term implications associated. The first generation of a true global bus-standard in the market was CAN. When introduced it was mainly considered for the interconnect between electronic modules of fairly big size: e.g. connection between motormanagement, gearbox, breaking-unit, dashboard among many others. LIN has now created the basis to introduce a second level of bus structures to link small periphery cheaply into the modules with lower datarate requirements. Since some years it has been recognized that standardization of interfaces helps to speed up the introduction of new component technologies. However this conflicts with fast introduction of innovative concepts, when the boundary of a component interface architecture is too small. The search for uniqueness versus the search for cost effectiveness dominates the discussion on electronic interface architectures.

2   Component Based Architecture

Today's car electronics architecture finds itself in boxes placed around the car. Typically these boxes are a result of the function definition, which has been generated during car design and reflects the organization structure of the car OEM. The reason for this is that the function definition is one-to-one translated into a workbook specification and into a piece of electronics hardware which in turn reflects the most cost effective solution within the definition perimeter.

Today's definition process is starting to differentiate between features and functions, when in the past this was not the case. The feature, which describes a certain characteristic of the car (can play radio, can control air temperature, can accelerate to 100 km/h in 8 seconds, consumes 6 litre per 100km) is translated into a function set and subsequently into a functions of electronic modules. However an overlap of features into several function domains is taking place. For this reason the system integration is becoming an ever increasing problem in car design. Only with a new architectural approach can this integration be mastered in the future. In Fig. 1 the function f_x belongs to both functional perimeters. Example: Dimming the sound of the radio when opening the door belongs to the perimeter radio as well as to the domain door.

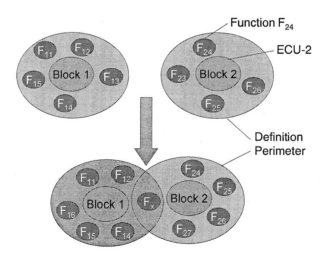

Fig. 1.   Function, Definition Perimeter and Module Relationship

The increasing number of functions in the car has led to an ever increasing number of electronic modules, which are approaching 100 in high end cars. It is predicted that this trend cannot be sustained, as the integration of an increasing number of modules into the overall car structure creates problems of unmanageable complexity. The complexity is due to several factors:

- The function boundaries are not distinct, but functions merge at the periphery, belonging to several definition perimeters.
- The coordination of a timely development of all modules simultaneously increase exponentially with the number of modules.
- The quest for uniqueness of new features makes the modules tailored to a specific application and reduces its total volume.
- The car designer's need for cost effective scalability adds to the problem, as the car is value priced in accordance to the functions implemented. Example: the car owner pays extra for a sun-roof, a seat-heater, a fold-function at the mirror or a rear seat passenger airbag. The basic, unequipped car however cannot bear any overhead cost associated with the capability to integrate these additional functions.

For this reason we have seen distributed architectures emerge in the automotive world, where functions and features are disjoint. This translates into solutions where the function and the control are not any more closely coupled, but work through a public interface accessible to other feature-controllers.

3   Bus Standards for Sensor Systems

There is a multitude of so called "Standards" which are available for sensor data transmission. A recent count from Christopher Lupini [1] showed about 43 busses proposed by different car makers and suppliers all relevant for the car. At this place the ones with potential for truly global distribution will be described.

As a generic overview it seems that the old classification of class-A (slow, up to 20 kb/s), class-B (medium 20-200 kb/s) and class-C (faster than 200 kb/s) is going to disappear and instead we see the following classification as being more relevant in the car.

3.5 Further Functions and Applications

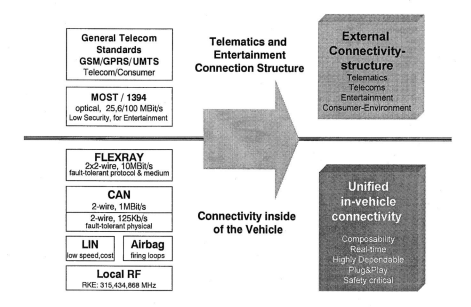

Fig. 2. In-vehicle Networking Development

Fig. 2 refers to the standards with largest distribution:

- Cheap subsystem interconnect up to 20 kb/s (UART-type) (LIN)
- Fault-tolerant medium speed used for interconnecting modules (Body-CAN)
- By-wire Chassis interconnect for safe systems (10-100 Mb/s) (FlexRay)
- Multimedia Interconnects (MOST, IEEE1394)
- Airbag (Safety) specialty connection
- Short range RF-Links

In the following sections the major standards and its compelling features should be described under consideration for usage of sensor integration.

3.1 LIN

LIN is quickly gaining momentum as a cheap interconnect for subsystems such as sensors or actuators. It was created to propel the Mechatronics concept, where the driving electronics is placed in close link with the actuator or sensor. Obviously this bus is holding the biggest potential to play the key role for future sensor busses because of a few compelling features:

- It is based on the broadly available SCI-hardware, which is basically a UART, asynchronous serial data transfer. This makes hardware very available.
- It introduces Time-Triggered principles and thus guarantees signal latency times on the bus, a feature which the well introduced CAN-bus is not able to perform without special additions.
- It does not need an external crystal and contends with internal trimmable RC-oscillators, thus making the setup extremely cheap.
- There are some clever power-saving and wake-up modes defined to save on current consumption of the system.
- The silicon is very cheap as it does not require an MCU in its simplest implementations.

Some disadvantages remain however:

- The data rate is limited to 20 kb/s due to single wire operation with 12 Volt swing. A higher data rates are not possible without introduction of massive EMC-problems.
- A second wire is necessary for power supply and in general a 3rd wire too for ground connection, bringing the total bus bundle to 3 wires.
- The LIN-bus is a pure Master-Slave bus. This means one master handles the complete traffic on the bus and is responsible for the schedule of signals to pass.

LIN is ideally suited to fit into a time-triggered global vehicle network. Together with FlexRay on the high-speed end of the spectrum, it can work to the pace of an in-vehicle master clock and deliver slow sensor signals cost effectively into the total in-vehicle network.

3.2 CAN

The CAN bus exists in two flavours:

- the low speed, fault tolerant version of 125 kb/s
- the high-speed version running typically at 500 kb/s

The maximum speed of the CAN bus is limited to 1 MB/s due to the arbitration scheme of a contention bus. The CAN bus is now well introduced in cars with the low speed version connecting all the body-modules and the high-speed version connecting the powertrain, gearbox and suspension.

In brief here the major advantages:

- CAN is widely introduced and has a lot of services established for it.
- CAN runs on a twisted pair with differential signal swing, so that the EMC-behaviour is good even for higher data rates up to 1 Mb/s.
- The fault-tolerant CAN is extremely versatile in handling bus failures with the ability to recover from faulty situations on the bus. This makes the bus extremely robust for situations of every day's life with vibrations, corrosion or humidity on the connectors.

Major disadvantages are:

- difficulties to compose the system, as new signals change the priority sequence, thus altering its latency times, huge interdependencies between bus-participants due to contention.
- The event driven bus system is making effective use of the bus bandwidth. However latency times are not guaranteed unless data traffic is limited through special means, bringing down the theoretic bus bandwidth significantly.

In comparison to LIN the CAN-bus is faster, thus able to manage more signals. However the demand for system resources is higher, demanding a good crystal based oscillator at each node. It needs one more wire than LIN, bringing the total wire count of a CAN-bundle to 4, when power and ground are included. Typically a CAN-node requires a microcontroller to perform the complex data processing functionalities the CAN-protocol requires.

For sensor or actuator signals however the CAN-bus remains a good fit as the CAN is in total well introduced and the number of CAN compatible devices is the largest, with about 100 Mio nodes produced in 2000, according to the CiA, the CAN-standardization body [6].

3.3 FlexRay

The next generation x-by-wire demands for a secure solution to interconnect the future chassis system, which will consist of the following major elements: Motormanagement, Gearbox-control, Suspension-control, Electric-Steering, Braking-control, Stability-control calculation, Sensor-signal cluster. To cope with the real-time constraints of the signal transmissions, only a time-triggered solution is acceptable. Today two solution are in development: TTP and FlexRay. FlexRay contains most of the time-triggered properties of TTP and in addition has the advantage to possess a protocol section where dynamic signal allocation according to signal priority is performed [7].

The FlexRay major features are:
- 2 channel communication (redundancy)
- each channel can be copper (twisted pair) or optics (fibre)
- bus guardian to prevent babbling idiot bus blockout conditions
- composability of the system due to time-triggered architecture
- dynamic signalling for control messages
- data rate of 10 Mb/s per channel expandable to 100 Mb/s and above

However the bus system is complex. It is well suited to ensure secure data transmission, but its price is well above LIN or CAN. Binding sensor-data from a sensor cluster into the system will have to be managed via gateway structures for cost-effective interfacing to lower-speed sensor data.

3.4 Multimedia Busses (MOST/IEEE1394)

Today the MOST-bus is making its way into the car in order to connect multimedia devices such as the CD- or DVD-player, the display-console, GPS and navigation, the telephone system and the back-seat entertainment displays. Although there are other multimedia standards in the consumer-electronics world such as IEEE1394 or USB, the car industry has chosen the MOST-standard, which is unique to the car-industry. It works on an optical ring with a 26.5 Mb/s data rate, expandable to higher data rates according to its roadmap.

Due to its purpose to link audio and video data over the bus, it needs to be considered for next generation image sensors. Imaging in the car will become a very important input signal for the following functions:

- driver detection (out of position) for safety (airbag) systems
- Driver viewing direction detection
- Car surround sensing (for self-parking or precrash detection)
- Lane-detection for lane-guided cruise control
- Car-to-car automatic signalling and many more

In order to link all these optical capture signals into the computational units, one of these connections will be required. Today the linkage over a bus is still a reach-out, as standard video formats from the computer world are being used.

3.5 Safety-Busses (Airbag)

Airbag has been a very dynamic field of investment in automotive over the past years and as a consequence traffic fatalities are on an all time low over the world. Due to its very competitive nature, the system providers have all invested into reducing the price of component connectivity. In consequence the number of solutions today is confusingly large without the hope to see a convergence into one final standard.

The major ones are: SafetyBus from Delphi, BST from Bosch-Siemens-Temic, DSI from Motorola, Planet form Phillips, and some more. This limits the attractiveness of the airbag component market to potential suppliers as available volumes are split. Some features are all similar for those bus structures:

- focus on very low cost
- single wire to pass data AND power, bringing the wire-bundle down to 2 wires
- consequential master-slave design
- no crystal at the slaves

With the exception of Byteflight, which is becoming a subset of FlexRay, today all these safety busses are not spreading beyond its usage in the safety (airbag) system. Unless this happens, the generic usage of these bus systems will remain limited.

3.6 Short range RF-links

For systems which cannot be connected with a wire, Radio Frequency transmission (RF) is playing an increasing role. This is the case for tyre pressure monitoring systems, for remote keyless entry, garage door opening, or even steering-wheel connections, thus only short range data transmissions, without using the transmission methods of the common telecom infrastructure. There are limitations associated to RF, which are mainly bandwidth and availability, as the air-interface is shared with other participants. There are limitations associated to RF, which are mainly bandwidth and availability, as the air-interface is shared with other participants.

Today the remote keyless entry modulation methods are available to transmit a few bytes of data with a few kilo-bit per second data rate. The RF-specification is changing form country to country, however the following standards have emerged on the following frequencies: 315MHz, 434MHz and 868MHz with FSK and OOK modulation. Integration of the sensor signals with the transmitter is not easy to achieve, monolithic integration is not on the roadmaps, so that the solutions today rely on multi-die packaging. In this context, the DC-Bus could be classified as an RF-system over the power network. In fact modulation and demodulation has to be performed and the power lines are being used as antennas.

4 Future System Setup

The next generation bus structures must allow a seamless integration of components. This requires a number of features to be available in the network system. Fig. 3 shows the concept of an all-available network structure, allowing new freedom in composability not realizable today. Some of the key features are listed below:

- **scalability**: the bus system's ability to integrate slow, cost effective nodes together with faster networks without adding to the baseline cost of the system. This allows to keep the cost of the electronic backbone in the car low, and add cost as much as new systems are being integrated.
- **faster networks**: higher data rates will be available in future systems. The advent of optical data transmissions will additionally boost throughput.
- **determinism**: schedulability of signals, synchronization and timeliness are increasingly demanded by the network systems to fulfil the timing constraints of the next generation real-time systems. Unlike PC-systems, where the user may wait and undefined time for completion, safety critical systems in the car need guaranteed response times.
- **failure** tolerance: error processing and treatment, special failure modes, voting and other prioritisation schemes will be critical to manage fault tolerant systems to a predetermined pattern. This will guarantee fall-back routines are available to make sure a next security level acts on the system to bring the car in a safe state when errors occur. This will require degradation policies to exist as a standard on the market.
- other **services**, such as variable bandwidth allocation, on-line diagnostics, hot swap and many more features must be present to setup a rugged field-serviceable structure.

5 Sensor Concepts for the New Network

Today Sensors are mainly equipped with a ratiometric output. In order to fit into the systems they need a bus interface. Many of the sensors are slow and would fit without problems into a bus structure such as LIN. The challenge is the integration of the bus interface which is only possible, if there is already a CMOS-logic in the system. In many cases this applies as the raw data signals are small and need to be conditioned by a data-processing unit.

The accelerometer of Fig. 3 is an example. The capacitive sensor element is built with micromachining technology and is the die on the left of the package. The signal processing of the output signals from the accelerometer cell is done on the right die in the same picture. As this is normal CMOS technology it is cost effective to add the LIN-bus interface on this die. Obviously a small cost adder will occur, as additional die size is required to perform the bus-interface function.

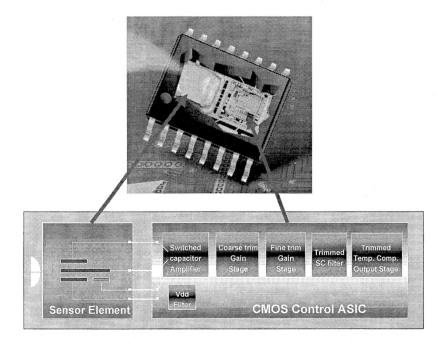

Fig. 3. Dual-die, Single Package Accelerometer

However this will be more than offset by the volume multiplication the standard interface should cause.

6 Reliability and Quality

Automotive Semiconductors have the best failure ratings compared with electronic devices used in consumer or computer applications. This is due to a large extent to the careful design for an extended temperature range and the rigorous qualification procedures for automotive semiconductors. Table 1 shows a requirements comparison. Especially the rigid QS9000 qualification requirements have helped to push automotive quality beyond consumer standards.

	Consumer and Computing	Automotive Electronics
Product Lifetime	3 Years	20+ Years
Innovation Cycle	6-9 Months	1.5-3 Years
Qualification	ISO9000	QS9000
Temp. Range	0°C - 70°C	-40°C to 125°C (150 °C)
Field Failure Goal	<100 ppm	< 1 ppm

Table 1. Automotive Quality Specification Comparison

Silicon sensors show the same excellent reliability as other silicon based solid state devices as long as their structure is protected from contact with the outside media. The available statistics show however that failure of electronics is to a large extent due to contact failures, especially the mechanical connectors exposed to the harsh automotive environment are the prime failure source in the vehicle. The automotive industry has understood these dependencies and is acting with appropriate qualification measures on connectors.

Of course will the integration of the next generation sensors into the automotive network help to improve quality further due to higher volumes and improved diagnostics.

7 Why Is It so Hard?

The industry lacks consensus that standardization is mandatory. The higher additional cost of the digital interface can only be offset over higher volumes. These however are not guaranteed as long as a standard is not established.

How can the industry overcome this threshold cost? Several opportunities exist:

- Drive industry consensus and mandate the bus-architecture starting by the car OEMs. This seems to be the obvious route and the telecoms industry shows how the early interface standardization process leads to quick industrial penetration results. (e.g. GSM interface definitions) However this requires a global consensus mechanism through standardization working groups, which the automotive industry is not entertaining on a large scale at this point in time.

- Have the sensor module makers team up to specify interface definitions: This way is most unlikely as the component suppliers are in fierce competition and use every differentiator to create a competitive difference rather than commonality.

- Introduce lifetime accounting: Rather than concentrating on the component price at the time of production, it would help to introduce some sort of lifetime-accounting. Considering the cost of the component over the whole lifetime starting from production, through service and spare part supply until recycling and scrap would introduce a new view and would work in favour of exchangeable interface standards.

In summary only through cooperative efforts of all industry partners will we be able to solve this pending problem. As the number of electronic components in cars will continue to increase, an agreed architecture based on well defined interfaces will be the only way forward for the entire industry. In fact the car industry is starting to recognize that it is beneficial to lead this process rather than wait and lag. As the time constants are slow, and the payback is long term over the lifetime of the car, the winners are those companies whose solutions become standard over the course of the years.

8 Conclusion

The application of agreed interface standards for automotive actuators and sensors is urgently required to allow faster technology introductions, lifetime design and availability, as well as ease of use to facilitate total systems integration. The challenge for the industry is how to converge the current bus structures into this visionary automotive backbone of the future for true composability of electronic components. The current microsystems product portfolios must incorporate existing interface standards to generate a truly open automotive platform for component plug&play in the rugged and highly reliable automotive environment.

References

[1] Christopher A. Lupini; "Multiplex Bus Progression" SAE-technical Paper 2001-1-0060, SAE2001 World Congress, Detroit Michigan.

[2] H.-C. von der Wense, A.J. Pohlmeyer; "Building Automotive LIN Applications" Advanced Microsystems for Automotive Applications 2001, Springer Verlag, ISBN 3-540-41809-1, pp279-292.

[3] G.Teepe; "New Architectures for faster Automotive Design Cycles" Advanced Microsystems for Automotive Applications 2001, Springer Verlag, ISBN 3-540-41809-1, pp121-130.

[4] G.Teepe; "Maßnahmen zur Steigerung der Innovationsgeschwindigkeit in der Automobilelektronik, ein Vergleich mit der Computerindustrie" Fachtagung Elektronik im Kraftfahrzeug in Baden-Baden, Oktober 2000, VDI-Berichte Nr. 1547, ISBN 3-18-091547-1, pp439-446.

[5] LIN-Consortium; "LIN-Specification Version 1.2" www.lin-subbus.org, Dezember 2002.

[6] Can in Automation (CIA) webpage on: www.can-cia.de

[7] Flexray Consortium webpage on: www.flexray-group.org

3.5.2 Plastic Packaging for Various Sensor Applications in the Automotive Industry

I. van Dommelen

European Semiconductor Assembly (Eurasem) BV
Microweg 1, P.O.Box 566, 6500AN Nijmegen, The Netherlands
Phone: +31/243714497, Fax: +31/243770406
E-mail: i.vandommelen@eurasem.com, URL: www.eurasem.com

Keywords: assembly, plastic packaging, optical sensors

Abstract

The application of integrated sensors in the automotive market is increasing. More than with packages for standard IC's, packages for sensor chips have a significant contribution to the functionality of the final electronic device. Besides the protecting function, the package should allow the sensing medium to have access to the chip surface.

Eurasem has developed and patented a special moulding technology, which makes it possible to leave a part of the (sensor) chip free of duroplast moulding compound where the gold wires are encapsulated as in a standard plastic IC. The advantage of this technique compared to other solutions is that the gold wire to bond pad interconnection has the same reliability as in standard plastic IC-packages, which is widely applied for automotive applications.

1 Introduction

Eurasem is an assembly subcontractor for plastic semiconductor packages in the Netherlands (Nijmegen). Besides offering standard plastic packages (QFP's, SO's) Eurasem is focusing on a niche market for (customised) special plastic packages. In January this year Eurasem has been acquired by Elmos Semiconductor AG, a well-known supplier of semiconductors for the automotive market. Although Elmos will supply Eurasem with a basic assembly load, Eurasem will continue operating on the open market as an independent assembly subcontractor offering both standard and special packages.

With the increasing application of sensor chips in the consumer-, and specifically in the automotive-market, packaging has become a major issue. For sensor chips that need a physical contact with the outer world like – optical-, humidity-, pressure-, flow-sensors there is one main topic. On one hand the packaging should protect the chip against both mechanical damage and hazardous environmental conditions, where on the other hand the sensor surface should have access to the medium to be sensed. To achieve this the sensor area of the chip needs to be exposed to the 'outer world'. The Eurasem cavity technique offers this possibility and has some specific advantage above other packaging solutions.

2 The Eurasem Cavity Package

2.1 The Basic Manufacturing Flow

The manufacturing of the cavity package is based on the flow of a standard plastic package assembly as is shown in Fig. 1, only the moulding process has been modified.

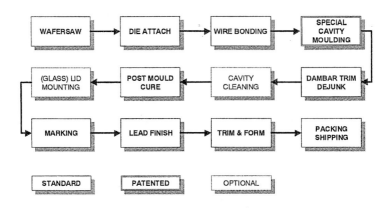

Fig. 1. Schematic Drawing of the Flow

During the moulding process the sensor area is kept open. This is achieved by a special nipple, which is gently pressing on the silicon chip surface, during the moulding process. When the fluid compound is pressed into the cavity (as in a normal transfer moulding process) the nipple prevents the moulding compound from reaching the sensor active area. In this patented concept the gold wires are covered with moulding compound during the moulding process as in a normal plastic package. As can be seen in Fig. 2.

3.5.2 Plastic Packaging for Various Sensor Applications   291

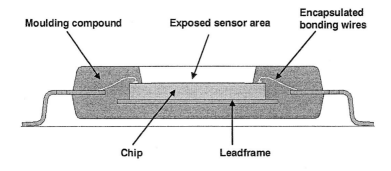

Fig. 2.   Cross Section of Cavity Packaging Concept

The special nipple design and material prevents the active area from being damaged by this physical contact. This special moulding technique can be applied on any plastic package type and depending on the application area, the cavity can optionally be covered with a lid. For optical applications this can be a glass lid or optical filter and for pressure sensor this could be a tube adapter. For every application a chip /sensor area specific mould half needs to be designed. Further backend processing remains unchanged.

2.2   Chip Design Guidelines

In order to be able to apply this packaging technique the chip design should meet some guidelines. In order to prevent the nipple from touching and damaging the wires there should be a certain distance between the sensor area to be exposed and the bond pads. This is further clarified in Fig. 3.

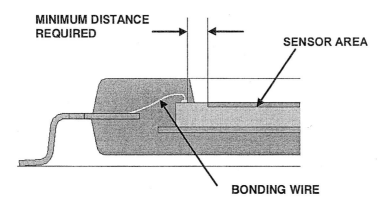

Fig. 3.   Design Rule Clarification

This minimum distance is determined by number of processing factors, to mention some:

- Die to lead frame placement accuracy
- Lead frame manufacturing accuracy
- Lead frame to mould alignment.
- Nipple manufacturing accuracy
- Minimum encapsulation needed for protection of the wire.

The minimum distance needed in the current process is 0.35 mm for optical applications. For pressure and flow sensors the minimum distance needed is dependent from the design and sensitivity of the sensor surface; a typical value is 0.50 mm. This spacing area can be used for additional electronic circuitry like data processing etc.

2.3 Further Processing

As can be seen in the basic flowchart of Fig. 1, the backend processes are standard. For optical sensor application some additional processes have been developed. The first is cleaning, in this process step which is just prior to glass lid mounting the optical sensitive area is cleaned in order to remove foreign particles which could block optical pixels. After cleaning the cavity is closed with a glass lid. A lot of development was put in this lid mounting operation; It was very difficult to attach the glass lid in such a way that it could withstand the high soldering temperatures. At the

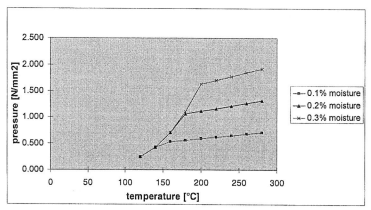

Fig. 4. Pressure Inside Cavity (2.5x 2.5 mm) Versus Temperature for Various Moisture Levels

soldering temperature (> 220 °C) the moisture which is present in the moulding compound will start boiling. From modelling it was calculated that this can cause a pressure inside the cavity of up to 10 bar. One of the main design aspects in this respect is the moisture uptake by the compound. The graph of Fig. 4 shows the predicted pressure increase versus temperature for various compound moisture saturation levels.

Currently the glass lid package can withstand MSL level 3, with the limited temperature of 220°C +0/−5°C. Further development is ongoing to improve the lid seal temperature resistance to the newest (lead free) soldering temperature requirement of up to 260°C.

2.4 Advantages of Cavity Package

There are different packaging alternatives for sensor applications available for example; - ceramic, - metal can, - pre-moulded, transparent (optical), etc... These concepts have one aspect in common. In order to give access to the medium the complete chip surface is exposed to the environment, sometimes with some medium transferring material like gel for pressure sensor application. Depending on the application area the Eurasem cavity technology (which only exposes the sensor area) has some significant advantages above the alternatives:

- The encapsulated bonding wires assure the same reliability in harsh environment as is achieved in proven standard plastic packaging technologies. This is a major advantage for direct contact sensor devices like fingerprint- (non-optical), flow- and chemical-sensors.
- It furthermore prevents scattering of light on the bonding wires for optical applications.
- The selective cavity opening enables other parts of silicon to remain unexposed. This can be used for dark current compensation.
- For optical application the air filled cavity automatically compensates for the light refraction of the glass lid. It enables this technique to be applied on small pixel image sensors using μ-lenses
- Another not unimportant aspect is cost. The Eurasem plastic cavity package is less expensive than the before mentioned ceramic and metal-can packaging alternatives.

3 Examples of Applications

In the Table 1 below different package types for which this cavity technology is available are shown.

Type	JEDEC reference	Cavity Size	Chip Size
MQFP 44, 10x10x2.0	MO-112-AA-1	4.35 mm SQ	6.7 mm SQ
MQFP 52, 10x10x2.0	MO-112-AC-1	4.35 mm SQ	6.7 mm SQ
PLCC 44	MO-047-AC	7.9 mm SQ	8.8 mm SQ
GLP-5 Linear Array	Customised Package	2.5 x 9.5 mm	8.9 x 1.2 mm
SMD-8 Linear Array	Non-Jedec	2.5 x 9.5 mm	8.9 x 1.2 mm

Table 1. Different Package Types for the Available Cavity Technology

All of the above packages were developed for optical applications, but can be applied for other applications as well. Feasibility samples were made with a flow sensor (1 µm) in the MQFP 44 package see Fig. 5 below.

Fig. 5. MQFP 44 Cavity Package Applied on Flow Sensor

Currently developments are ongoing regarding a SOIC 8 package for an oil condition sensor application. The cavity size is 2.5 x 2.5 mm. This package will be applied directly in contact with motor oil at a temperature up to 140 °C. First feasibility samples for this application in the above SMD-8 package (without glass lid) show promising results.

4 Packaging Reliability Aspects

4.1 Qualification Tests

The before mentioned GLP-5 linear array package has been qualified according a standard automotive qualification program (summary Table 2).

Package Size: 7.6 x 15.2 x 2.65 mm
Number of I/O's: 5
Cavity Size: 2.5 x 9.5 mm
Glass Lid Dimension: 12.5 x 5.5 mm
Die Dimension: 1.2 x 8.9 mm

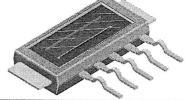

Test	Condition	Duration	Failed
Temperature Cycling	-65 → 150°C	500x	0 / 120
Thermal Shock (Air to Air)	-40 → 125°C	1000x	0 / 77
High Temperature Storage	150°C	1000 hrs	0 / 48
Cycled Temperature Humidity Bias	90-85%, 30–65°C	1000 hrs	0 / 77
Salt Atmosphere	TM1009		0 / 15
Gross Leak Test	TM1014 Condition C1		0 / 15
JEDEC MSL 3	JESD-22-A113-B		0 / 77
Accelerated Lifetime Test	85°C/ 85% RH	1000 hrs	0 / 48
Sequential Mechanical Test	TM2002 cond. B, 1500g TM2007 cond. A, 20g 20 Hz-2KHz		0 / 30

Table 2. Summary

4.2 Moisture Induced Test Considerations

A reconsideration on the moisture induced qualification tests (HAST, CTHB, 85% / 85°C) for the optical variant of the cavity package is needed. Moisture does not only have influence on the interconnection reliability, it has also a direct influence on the optical functionality.

The duroplast package is non-hermetic and therefore permeable for moisture. With as a result that if a plastic cavity package with lid is exposed to a certain environment like moisture, the moisture will penetrate into the cavity independent from the lid seal integrity. This moisture in the cavity can eventually cause condensation when the package is exposed to low temperatures immediately. This seems to be a drawback from this concept when a (glass) lid is applied, however the following has to be taken into account:

- Acceleration of corrosion of the interface between (gold) wires and the (aluminium) bond pads is the target of all moisture induced tests. It is not a mirror of real system environmental conditions.
- The condensation effect after current test will also take place on the second level packaging like lenses, thermoplastic housings etc. An improved first level package would therefore bring no overall system improvement at all.

Alternative ideas to overcome this problem:
- Skip the application of a lid.
- Deliberately make the lid seal not gross leak tight but apply it only as mechanical protection of the sensor surface. In that case the moisture can get in and out more rapidly, decreasing the risk for condensation.

4.3 IR-reflow Soldering Temperature Resistance

Some remarks to soldering temperature resistance of electronic components in general. There is nothing against the initiative to reduce the amount of environmental harmful materials like Lead. It is obvious that lead-free plating require a higher soldering temperature. On the other hand, it should be considered whether it is wise to design all electronic components for a very severe condition which they are only exposed to just a very small fraction of their lifetime (maximum 3 x 10 seconds over 10 - 15 years). Effort should be put in reducing or not increasing the soldering temperature for example by proper solder pastes.

5 Conclusions

With the development of the special moulding technique, Eurasem succeeded in applying a unique duroplast plastic packaging concept for sensor applications. An optical linear array sensor package has been qualified for the automotive industry and in running production now. In order to be able to apply this technique the sensor chip has to be designed according some guidelines.

6 Acknowledgements

I would like to thank the development team of Eurasem for the amount of effort put into this special packaging technique development, especially my colleague Jurgen Raben.

3.5.3 Harsh Fluid Resistant Silicone Encapsulants for the Automotive Industry

K. L. Pearce, E. M. Walker, J. Luo, R. A. Schultz

Emerson & Cuming
46 Manning Road, Billerica, MA 01810, USA
Phone: +1/978/436-9700, Fax: +1/978/436-9707
E-mail: kate.l.pearce@nstarch.com, URL: www.emerson&cuming.com

Keywords: silicone, fluorosilicone, gel, encapsulant, circuit protection

Abstract

Silicone gels have found widespread use in automotive electronic applications as encapsulants and potting compounds for the protection of electronic devices. The continued decrease in size of electronic packages is enabling the placement of sensors closer to the point of measurement, for example under hood, on exhaust manifold, on wheels, on brakes etc. This is resulting in the necessity for protection materials that can withstand increasingly harsh environments: higher service temperatures and harsh chemicals, for example fuels, oils, acids and other aggressive automotive fluids. Silicone gels are known to protect sensitive electronics from mechanical and thermal stresses as well as dust and dirt. Fluorosilicones provide increased resistance to high operating temperatures and fluids such as fuels. Often these materials cannot provide protection from the diverse range of conditions typically encountered in these applications. Such conditions may include exposure to fuel and acids. Materials may fail due to polymer degradation, bulk swelling which allows corrosive fluids to permeate them, or more likely because the interface between the protective material and the package has failed.

This paper will describe a material based on a proprietary silicon hydrocarbon hybrid resin which offers excellent protection from harsh fluids including fuel and acids. The material is a soft elastomer with improved strength and toughness compared to similar materials based solely on silicones or fluorosilicones. The material demonstrates good adhesion to a variety of substrates; on exposure to harsh fluids including fuels, acids and oils, the adhesion properties of the material are retained, and the material maintains its physical integrity. A second material set derived from this proprietary resin demonstrates superior strength and toughness compared to similar materials based solely on silicones. The physical integrity of the second material set retains its adhesive properties on exposure to harsh fluids including alcohol-based fuels and acids. Both materials described in this paper are suitable for providing an increased level of protection to electronic packages, which may be exposed to harsh fluids.

1 Introduction

Since the introduction of the first automotive sensor, many applications have evolved based on the variables that it is now possible to measure. Table 1 gives examples of the capabilities of modern day sensors.

Variable	Application
Pressure	Inlet manifold, Oil, Tire, EGR, Atmospheric, brake fluid, turbo boost, fuel tank vapor, injection pressure
Temperature	Coolant, interior air, external air, seats, engine inlet air, exhaust gas
Flow	Inlet air mass, fuel flow
Speed	Wheel speed, engine speed
Position	Crankshaft, camshaft, throttle, accelerator pedal, axle (height), steering wheel, suspension
Level	Fuel, washer fluid, engine oil, radiator coolant
Oxygen	Concentration in exhaust
Shock	Engine knocking
Yaw Rate	Vehicle yaw (understeer and oversteer)
Acceleration	Vehicle impact, lateral acceleration

Table 1. Examples of Automotive Sensor Applications

Automotive sensors have also increased in sophistication such that engine management systems are now able to combine several functions to provide overall engine and emission control. Examples of such sensors include mass airflow sensors, MAP (pressure) sensors, fuel tank vapour pressure sensors and exhaust gas temperature sensors. The driver of such applications is often increasingly stringent regulations. Furthermore sensor utilisation has led to "smart systems", capable of aiding occupant protection and accident avoidance. These may indicate the presence, position and weight of occupants, detect relative wheel speeds for traction and chassis control, detect rain, moisture and light level for automated wipers and headlamps and detect proximity to obstacles for the purpose of collision avoidance or as a parking aid.

A primary enabling factor in the advancing technology of automotive sensors is the decreasing size of electronic packages resulting in the

3.5.3 Harsh Fluid Resistant Silicone Encapsulants for the Automotive Industry

capability of placing sensors closer to the point of measurement. A repercussion of this capability is the necessity for automotive sensors to withstand higher service temperatures and increasingly harsh environments. Thus while automotive sensors which are to be located in the cabin of a vehicle may need to withstand relatively mild temperatures, sensors placed on the engine or the transmission of a vehicle are required to withstand far higher temperatures. Correspondingly, sensors falling into the latter category may also be subjected to harsh fluids including fuels and motor oils, transmission fluids as well as the degreasers used in engine cleaning processes. Fig. 1 and Table 2 summarise the temperature specifications and fluid tolerance that current automotive sensors are expected to function in.

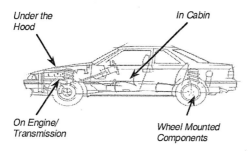

Fig. 1. Location of Sensors

Silicone gels are commonly used to protect sensitive electronics from mechanical and thermal stresses as well as from dust and dirt. A benefit of these materials is that they may be applied as a flexible coating to sensors without adversely affecting their performance; they also show excellent resistance to elevated temperatures. A limitation of silicone gels is that they do not commonly provide resistance to harsh fluids such as fuels, oils and acids. Typically the fluid penetrates their bulk or the interface between the protective gel and the electronic package resulting in polymer degradation or swelling, loss of polymer integrity and ultimately corrosion followed by failure of the electronic device. Fluorosilicones are known to provide protection from some fluids, such as fuels. However, often these materials can not provide protection from the combinations of conditions described in Table 2.

Component Location	Operating Temperature	Resistance to Fluids
In Cabin	-40°C to 90°C	Interior cleaner, Coffee, Soda, Alcohol, Fuels, Water/Humidity
Under the Hood	-40°C to 150°C	Coolant, Motor oil, Brake fluid, transmission fluid, Fuels, washer fluid, battery acid
On Engine/Transmission	-40°C to 180°C	Motor oil, transmission fluid, fuels, degreaser
Wheel Mounted Components	-40°C to 250°C	Hydraulic oil, brake fluid, salt spray

Table 2. Higher Service Temperatures and Harsh Fluid Environments

This paper describes new materials, which offer increased resistance to the fluids that automotive sensors are commonly exposed to, while maintaining the heat resistance also required for these applications. In addition, methods for screening materials for their suitability in such harsh environments will be discussed. The materials described are particularly suitable for on engine applications, for example with MAP sensors and transmission sensors.

2 Material Development

Two novel materials have been developed: both are silicone-based, one of which is fluorinated. These silicones are used in conjunction with a National Starch and Chemical (NSC) proprietary resin with the aim of improving the properties of the silicones. The proprietary resin is a silicon-hydrocarbon hybrid that demonstrates desirable properties including low moisture uptake, good electrical properties, adhesion, chemical resistance and physical toughness. It is expected that when combined with a silicone these properties will be imparted to the resulting material, which is expected to retain the thermal stability and flexibility over a wide range of temperatures normally shown by silicones. In addition, good solvent resistance is expected when the NSC proprietary resin is used in combination with a fluorosilicone.

3 Test Methods

The ultimate test of performance for an automotive sensor protection material is a field evaluation. Prior to this, automotive sensor manufacturers generally subject new materials to simulation tests that utilise the specific electronic components for which the material is being

3.5.3 Harsh Fluid Resistant Silicone Encapsulants for the Automotive Industry

considered. A materials formulator, who needs to screen numerous materials for their suitability in a range of environments in a limited time frame, needs to adopt a more basic screening test. Typically, formulators have published the results of bulk material testing, including the measurement of dimensional changes (volume swelling) or weight loss when a bulk specimen of the material under consideration is soaked in a particular fluid. Such evaluations give a good indication of the behaviour of the material in the fluid. However, in order for a material to provide adequate protection to a sensor device, it must be capable of forming an adhesive bond to the device and maintaining that bond on exposure to the fluid. Adhesion is an interfacial property that is substrate specific. The adhesion of a particular material may vary when the material is in contact with, for example, an engineering plastic, such as PPS (polyphenylene sulphide), a ceramic or a metal such as gold.

A method for evaluating the adhesion of the circuit protection material to a substrate involves application of the circuit protection material to the substrate and embedding reinforcing glass cloth into the material prior to curing. The glass cloth is coated with a suitable primer that will ensure adhesion of the circuit protection material to the cloth. After curing, the force required to peel the circuit protection material from the substrate at an angle of 90° is measured using a tensiometer. The cloth is used to reinforce the circuit protection material and provide tensile rigidity. The whole assembly: substrate, cloth and circuit protection material is soaked in a suitable test fluid, thus the effect of the test fluid on the adhesion of the circuit protection material to the substrate can be evaluated. The method is shown schematically in Fig. 2.

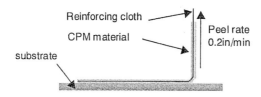

Fig. 2. Adhesion Test (Modified from ASTM D-413-82)

Materials were also evaluated for their tear strength, which is a measurement of the cohesive strength of the material. The method is based on the adhesion test described above, except that a strip of Teflon film is inserted between the substrate and the circuit protection material to prevent adhesion to the substrate. The force required to tear the cloth through the cured circuit protection material is measured using a tensiometer. Other test methods utilised include the measurement of tensile strength and elongation and the determination of the material hardness. These tests were carried out according to ASTM D2240 and ASTM D412 respectively.

Fluid immersion soak tests, described by ASTM D471 are described in Table 3.

Condition	Temperature (°C)	Time (hours)
Dry heat	180	varied
Fuel C	25	150
Fuel CM 85	25	150
Nitric acid, pH 1.6	85	400
Sulphuric acid, pH 1.6	85	400
Mobil 1 (used)	140	400
Transmission fluid	140	200

Fuel C is a 1:1 mixture of iso-octane and toluene
Fuel CM 85 is Fuel C mixed with 85% methanol

Table 3. Fluid Soak Schedule

4 Material Properties

The materials were tested initially for their hardness, tear strength, adhesion to PPS and gold and tensile strength/elongation. Data is shown in Table 4.

Property	NSC proprietary resin with fluorosilicone	NSC proprietary resin with silicone
Hardness (Shore 00)	65	57
Tear Strength (Jm^{-2})	800	743
Adhesion to PPS (Jm^{-2})	149	121
Adhesion to Gold (Jm^{-2})	175	86
Tensile Strength (psi)	56	130
Elongation (%)	420	285

Table 4. Performance Prior to Fluid Soak

The test specimens were then subjected to the conditioning schedules detailed in Table 3 and re-tested. The data showing the performance retention after conditioning are shown in Table 5 and Fig. 3 and 4.

3.5.3 Harsh Fluid Resistant Silicone Encapsulants for the Automotive Industry

Table 5 shows the hardening effect of heat ageing the materials at 180°C. As can clearly be seen, the NSC proprietary resin in conjunction with fluorosilicone shows minimal heat hardening. The majority of the hardening effect that can be observed occurs during the initial stages of the experiment and the hardness then levels out to a constant value during further testing. The NSC proprietary resin in conjunction with the silicone resin shows some hardening effect due to heat; improvements in this area are the subject of on-going research.

Time (hours)	Increase in hardness (%)	
	NSC proprietary resin with fluorosilicone	NSC proprietary resin with silicone
100	4.6	22.8
200	4.6	31.5
340	6.0	-

Table 5. Hardness After Heat Ageing

Fig. 3 shows the effect of fluid soaks on tear strength data. This test is primarily a measure of cohesive adhesion. The NSC proprietary resin in conjunction with fluorosilicone performs well in all fluids tested. The NSC proprietary resin in conjunction with silicone performs well in the fluids tested with the exception of Mobil 1, where some loss of tear strength is observed.

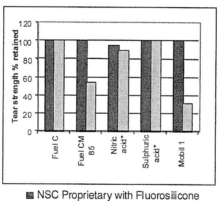

■ NSC Proprietary with Fluorosilicone
□ NSC Proprietary with Silicone

Fig. 3. Tear Strength Retention After Fluid Soak

Fig. 4 shows the effect of fluid soaks on the adhesion of the materials to both PPS and gold. Again good retention of properties is observed for the NSC proprietary resin in conjunction with fluorosilicone. Some loss of adhesion to PPS is observed for the NSC proprietary resin in conjunction with silicone on exposure to Mobil 1.

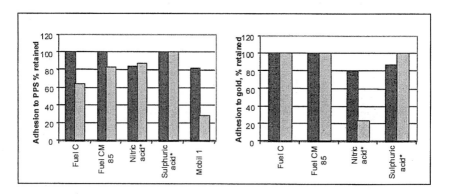

■ NSC Proprietary with Fluorosilicone
□ NSC Proprietary with Silicone

Fig. 4. Adhesion to PPS / Gold Retention After Fluid Soak

Fig. 5 shows the effect on the tensile strength and elongation of soaking the materials in various fluids. The NSC proprietary resin in conjunction with fluorosilicone resin shows no loss of tensile strength or elongation in the fluids studied. The NSC proprietary resin in conjunction with silicone resin shows good retention of tensile strength and elongation when conditioned in Fuel CM 85 and acid.

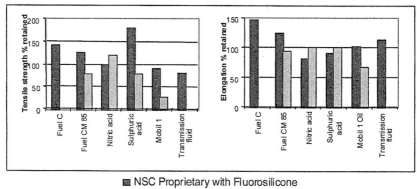

■ NSC Proprietary with Fluorosilicone
□ NSC Proprietary with Silicone

Fig. 5. Tensile Strength / Elongation Retention After Fluid Soak

5 Conclusions

Two materials have been described which show good resistance to the influence of heat and to harsh fluids. The NSC proprietary material in conjunction with fluorosilicone is a gel material that has demonstrated the good heat stability expected of silicone gels and the resistance to fuels expected of fluorosilicones. Furthermore it has demonstrated good resistance to more polar fluids, including alcohol based fuel and acids, a property expected from the NSC proprietary resin. The NSC proprietary material in conjunction with silicone is a soft elastomer that also shows good resistance to alcohol based fuels and acids. It has been noted that this material hardens significantly under the effect of increased temperatures and the correction of this problem is the subject of further research.

Acknowledgements

The authors would like to thank Mr. Charles Mooney and Mr Robert Palmer of Emerson and Cuming, Dr Nikola Nikolic, Dr Donald Herr, Mr Larry Scotchie and Mr Michael Cipullo of National Starch & Chemical for their guidance and help during this project.

References

[1] Reed Electronics Research, Automotive Sensors, Ouadrant House, 1998-2000.

[2] Norton, Sensor & Analyser Handbook, Prentice-Hall 1982.

3.5.4 Silicon-On-Insulator (SOI) Solutions for Automotive Applications

L. Demeûs[1], P. Delatte[1], G. Picun[1], V. Dessard[1], Pr. D. Flandre[2]

[1]CISSOID S.A., The SOI Design Specialist
Place des Sciences 4 bte 7, 1348 Louvain-la-Neuve, Belgium
Phone: +32/10/489210, Fax: +32/10/489219
E-mail: laurent.demeus@cissoid.com

[2]Microelectronics Laboratory, Université catholique de Louvain
Place du Levant 3, 1348 Louvain-la-Neuve, Belgium

Keywords: high temperature sensors, Low-Power, SOI, reliability, harsh environment, sensors

Abstract

This paper will focus on the use of SOI (Silicon-On-Insulator) technology for automotive applications. SOI has been recognized since a long time for its advantages for High Temperature electronics, low-power electronics and since some years, for MEMS applications. These three fields fit all future automotive requirements for electronics products.

1   Introduction

Silicon-On-Insulator (SOI) is an emerging semiconductor fabrication technology. SOI is already well known for its advantages for high speed application but the Automotive sector can also take advantage of this new technology. Two major needs in electronics systems for automotives can be reached with SOI technology, either High Temperature operation, or low-power consumption, to an extend depending on the exact used SOI process. MEMS applications can also take significant advantage for micromachining on SOI substrates. This paper will mainly concentrate on High temperature application.

The basic idea of SOI technology is to isolate the mechanical silicon substrate of a wafer from the active silicon layer (See figure 1). This yields to a number of advantages: less leakage currents, less parasitic capacitances, no latch up, better sub threshold slope, intrinsic isolation between transistors, no wells, better radiation resistance, ...

3.5.4 Silicon-On-Insulator (SOI) Solutions for Automotive Applications

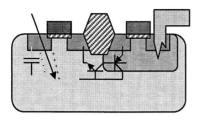 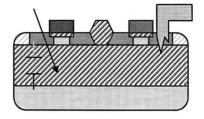

Fig. 1. Comparison of Cross Section of BULK (Left) and SOI (Right) NMOS and PMOS

SOI technology is a generic name for Silicon processes using SOI wafers, but each SOI process has its own specifications. The first distinction we can make is between thick (about 1 μm of active silicon) and thin (between 50 and 200 nm of active silicon) SOI technologies. Thick SOI requires additional trench isolation between transistors and is mainly use when vertical devices are required, this is not the subject of this paper. Thin SOI can only use lateral devices and is divided in two main families: Fully Depleted (FD) and Partially Depleted (PD) SOI, depending on the depth of the depletion region compared to the thickness of the active silicon layer. The most advantageous processes are Fully Depleted but are the most difficult to develop. The majority of processes available at this moment are Partially depleted processes.

As a conclusion, depending on the application and its requirement, we have to choose the best-suited SOI technology. But even if a SOI technology is optimised for one advantage, for example High Temperature, it will still present other advantages, for example, lower power consumption than a BULK technology.

2 High Temperature Operation

Electronics in Automotive applications moves to hotter localizations, close to engine, wheels and exhaust system. Operating temperatures for these electronics components far exceed the 150-175°C maximum rating of standard devices.

Since last year, High Temperature SOI technology for industrial applications is available in Europe and open the way to design highly reliable electronics circuits for years of operation up to 225°C. This technology is a Partially Depleted SOI process, optimised for High Temperature operation, including a specific High Temperature metalisation to avoid early electromigration. Fully depleted SOI processes are also available but only for research purpose.

The comparison of behaviour between classical BULK, PD and FD is summarized in the following Table 1.

	BULK	PD SOI	FD SOI
ΔVth (mV/°K) (Threshold voltage)	2-5	2.5-3	0.7-1.5
Ileak (A/μm) @ 300°C	1 μ-10 μ	10 n-100 n	1 n-10 n
1M SRAM standby power consumption (extrapolation @ 200°C)	0.5 W	20 mW	10 mW
1M SRAM standby power consumption (extrapolation @ 300°C)	>10 W	1 W	100 mW

Table 1. Comparison of Behaviour Between Classical BULK, PD and FD

We can easily see that Fully Depleted is the best suited process for High Temperature operation but it suffer from a high development cost and the difficulty to make High Voltage devices, due to the very thin silicon layer. In the other hand, Partially Depleted SOI is already very advantageous compared to Bulk technology and High Voltage devices, up to 150V (low current) can be processed in the silicon thickness. This is a tremendous advantage for automotive application to be able to sustain the future 42V power supply.

High temperature BULK circuits up to 200°C have been published since years but their design is very difficult and the silicon surface is largely increased compared to the same function at room temperature. SOI technology is now mature to propose an alternative to these BULK solutions. Even if some efforts has to be done to design these circuits on SOI technology, an experienced team in High Temperature SOI can achieve state of the art circuits in a very short time with small die area which can lead to competitive products for High Temperature market.

A lot of demonstrators have been published by the Université catholique de Louvain [1] on a fully depleted SOI process for research. But we will here focus on two new examples of High Temperature circuits for commercial applications designed on the PD SOI mentioned above. We summarize the results of these demonstrators below.

The first demonstrator is a Voltage regulator:

        Output Voltage: 5 discrete values between 3.3 V to 15 V
        Input Voltage: min Vnominal +1 V and max 40 V
        Output Current: 1A max @ 225°C
        Precision: 2% on absolute value and 2% of drift on t° range (-25; 225°C)
        Input ripple rejection: -60 dB @ 1kHz
        Die Area: 15 mm², 3 pins package

3.5.4 Silicon-On-Insulator (SOI) Solutions for Automotive Applications

The second demonstrator is an Instrumentatio Amplifier:

Differential input and output,
- Die area: 8mm²
- Power supply Voltage: 7V
- Fixed gain: 100
- Precision: 3/100 from 0Hz to 2kHz) on t° range: 0°C to 200°C
- Power consumption: 640μA @25°C and 2mA@ 200°C

These demonstrators are expected to be tested in march 2002 and produce for commercial availability at the end 2002. These circuits are expected to work 5 years at 225°C.

3   Low-Power Electronics

As the number of electronics components in cars increases each year, a new problem of total power consumption appears. That problem has lead automotive industry to begin a move from 14 V to 42 V battery levels. Beyond this effort, new electronics for automotive have to be low-power consuming. SOI can offer solutions for critical applications, where the electronics keep running when engine is off (remote receiver, positioning system, ...) or where electronics has to be battery operated or remotely powered as in wheel pressure sensors.

SOI technology presents a tremendous advantage for low-power consumption. A lot of papers have been presented these last years by processors manufacturer who are expected to move to SOI in the next 2 years. Classical figures published are an advantage of 30% in power consumption in addition to a 30% increase in speed.

Up to know, main results have been achieved by processor manufacturer but a lot of foundries are now working on deep sub-micron SOI technologies, either Fully Depleted or Partially Depleted. These technologies will be available in the next 2 years for ASIC design but also DSP and FPGAs.

In conjunction to low-power consumption for digital parts, SOI presents also a better frequency behaviour and an intrinsic isolation between analog and digital parts which decreases the crosstalk. Moreover, at high frequency, the isolation between the mechanical substrate and active silicon enable the use of a "high resistive" substrate to increase the quality of integrated passive components. This yield to a technically and economically competitive SOI technology compared to BiCMOS technology.

Automotive is a very promising market for SOI RF solution for short range communication as it don't require specific power amplifier and will enable the co-integration of digital and analog parts in a "one-chip" solution.

4   SOI for MEMS

Concerning the fabrication of sensors and actuators on membranes, such as in pressure sensors, micro-heaters for gas sensors, ..., SOI substrates may finally offer a very interesting potential, since the buried oxide intrinsically provides a very well-defined etch-stop for micromachining, as well as inherent decoupling and isolation of microstructures and semiconductor devices, which significantly eases the compatibility and co-integration of microsystems with CMOS circuits.

5   Conclusion

SOI technology is now available for High Temperature electronics in Europe and automotive represents a huge part of this market. According to HITEN, HT electronics for automotive will represent 64% of this market in 2008 ($887.1m). And SOI technology will certainly take the biggest part of this market. But SOI will also be used for its low-power ability and MEMS fabrication easiness.

References

[1] L. Demeûs, et al., "Integrated Sensor and Electronic Circuits in Fully Depleted SOI Technology for High Temperature Applications", IEEE Trans. On Industrial Electronics, vol 48 number 2, April 2001, pp.281-285.

3.5.5 Micromachined Gyroscope in Silicon-On-Insulator (SOI) Technology

A. Gaisser[1], J. Frech[1], G. Hoelzer[3], P. Nommensen[1], M. Braxmaier[1], T. Link[1], W. Geiger[1], U. Schwarz[3], H. Sandmaier[1,2], W. Lang[1]

[1]Hahn-Schickard-Gesellschaft
 Institute of Micromachining and Information Technology, HSG-IMIT
 Wilhelm-Schickard-Strasse 10, 78052 Villingen-Schwenningen, Germany
 Phone: +49/7721/943-226, Fax: +49/7721/943-210
 E-mail: alexander.gaisser@hsg-imit.de

[2]IZFM, University of Stuttgart
 Breitscheidstrasse 2b, 70174 Stuttgart, Germany

[3]X-FAB Semiconductor Foundries AG
 Haarbergstrasse.67, 99097 Erfurt, Germany

Keywords: gyroscope, angular rate sensor, SOI-technology

Abstract

This paper focuses on the development of a low cost micromachined Corioles Vibrating Gyroscope (CVG) realized according to the decoupling principle DAVED-LL (**D**ecoupled **A**ngular **Ve**locity **D**etector) [6] where the two oscillation modes are of linear shape. By an optimised design the coupling of both oscillation modes can be suppressed, the so-called quadrature signal can be reduced significantly and thus the sensor performance can be improved to meet the demands and specifications of automotive applications. The DAVED-LL prototypes show a rms noise in the range of 0.05 °/s and a quadrature signal between 50 - 70 °/s.

1 Introduction

Low cost and high precision gyroscopes are used for advanced automotive safety and comfort systems, robotics (home robots, autonomous guided vehicles), virtual / augmented reality, people-to-people and people-to-device communication (gloves, helmets, and mobile phones), and medicine (surgical instruments). Due to the large market scope, various groups [1], [2], [3] are working on new designs, technologies and readout concepts for micromachined gyroscopes.

Keeping in focus the low cost factor, a clear tendency to use 'surface-micromachining-like processes' can be observed. As several specialists already remarked [4], [5] we stand on the verge of a second silicon revolution.

2 Theory and Design

Vibratory micromachined gyroscopes rely on a driven oscillation and a detection oscillation. Latter is driven by the Corioles forces due to an external angular rate in presence of the primary oscillation.

The decoupling principle DAVED [6] is briefly described with an schematic drawing of the LL-structure as seen in Fig. 1.

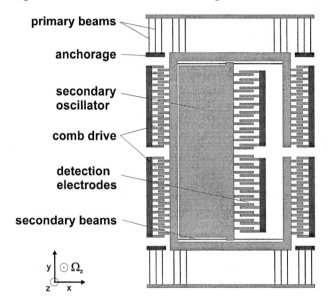

Fig. 1. Schematic Top View of the LL-structure (bright gray: Movable, black: Fixed).

The entire movable structure (printed in bright gray) is electrostatically driven to a linear oscillation parallel to the x-axis by comb drives (primary mode). When the device is turned around the z-axis Corioles forces excite an oscillation perpendicular to the primary mode, i.e. parallel to the y-axis. This secondary mode is detected by interdigitated electrodes and yields the output signal. Only the inner structure (secondary oscillator) can follow the Carioles forces. It is decoupled from the outer frame by a set of springs

3.5.5 Micromachined Gyroscope in Silicon-On-Insulator (SOI) Technology

(secondary beams). Due to the DAVED-LL design a better decoupling of the two vibration modes [7], [8] is achieved and the sensor performance is improved significantly. Thus, the secondary oscillation does not interfere with the driving oscillation and parasitic effects of the comb drives, like levitation, can be suppressed effectively.

3 Realization

A new, low-cost technology based on SOI-substrates and deep silicon etching (Fig. 2), was developed at HSG-IMIT and transferred to X-FAB.

Fig. 2. Schematic Cross Section of the Sensor Chip Realized with the New SOI-technology Developed at HSG-IMIT

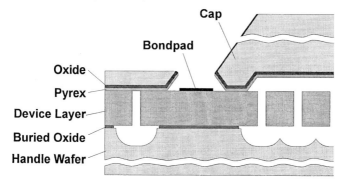

Fig.3. SEM-graph of the DAVED-LL Sensor Chip

Only two masks for the sensor wafer and two masks for the cap wafer are required for the fabrication of the complete sensor. Using this technology DAVED-LL sensors were fabricated (Fig. 3) and characterized.

4 Results

Parameter	Target specification	Pressure 10 hPa	Pressure 1000 hPa
Range:	±100 °/s	±100 °/s	±100 °/s
Resolution:	0.05...0.01 °/s	< 0.02 °/s	----
Bandwidth:	50 Hz	50 Hz	50 Hz
Noise: (RMS)	< 0.5 °/s	< 0.1 °/s	< 0.3 °/s
Bias drift *	< 2 °/s	< 0.1 °/s	< 0.4 °/s
Linearity **	< 0.5% (FSO)	< 0.1% (FSO)	< 0.1% (FSO)
g-sensitivity	< 2 °/s/g	< 0.2 °/s/g	< 0.2 °/s/g
Scale factor	10mV/°/s	10mV/°/s	10mV/°/s

* 0°C ... +70°C, ** End-Point-Straight-Line

Table 1. First results of the DAVED-LL Sensor

The target specification and results of first measurements of DAVED-LL gyroscopes produced by X-FAB are shown in Table 1.

5 Conclusions

In 1997 HSG-IMIT presented a concept for the design and the technology of a decoupled micromachined gyroscope. Within a sponsored project this new SOI-technology is successfully transferred to X-FAB, a silicon wafer foundry which will offers the necessary MEMS specific process steps. First samples of DAVED LL-sensors were fabricated at X-FAB and show promising results. It is believed that with the decoupled design a basic approach is established for the further improvement of micromachined gyroscopes. They can be fabricated with a simple low-cost surface micromachining-like process. By optimizing the design and the technology even higher performances for future gyroscopes are possible. An overall accuracy better than 0.1 °/s seems to be feasible and finally micromachined gyroscopes can reach a performance comparable with optical gyros but at a much lower price.

Acknowledgements

The work has been sponsored by the Federal Ministry of Education, Science, Research and Technology of Germany (BMBF) and by the Government of Baden-Württemberg.

References

[1] T.Matsuura, et al." A bulk Silicon Angular Rate Sensor using Micromachining Technology" , Symposium Gyro Technology 2000, Stuttgart, Germany , Sept. 2000.

[2] www.kvh.com/

[3] www.systron.com/

[4] Paul McWhorter-"Intelligent Microsystems; keys to the next silicon revolution", in MSTnews, No4, 1999.

[5] K.J. Gabriel – "Microelelectromechanical Systems", in Proc of the IEEE. vol. 86, No. 8Aug.1988.

[6] W.Geiger, et al., Patent DE 19641284 C1, May 20, 1998, HSG-IMIT.

[7] Y.W.Ilsu, et al., Patent US5955668, Sept. 21, 1999, Irvine Sensors Corporation.

[8] J.Green, Patent WO99/12002, Publication March 11, 1999, Analog Devices Inc.

3.5.6 Thermal Microsensor for the Detection of Side-Impacts in Vehicles

M. Arndt[1], R. Aidam[2]

Robert Bosch GmbH, Automotive Electronics Division
[1]AE/SPP5, Tuebingerstrasse 123, 72762 Reutlingen, Germany
E-mail: michael.arndt@de.bosch.com

[2]AE/ERP5, Postbox 300240, 70442 Stuttgart, Germany
E-mail: rolf.aidam@de.bosch.com

Keywords: side-airbag, side-impact, thermal microsensor

Abstract

Due to the small size of the lateral compression zone in automobiles, the response time for protective measures in case of side-impacts must be very short. Monitoring of the air temperature in the interior of the side doors with the objective to detect quasi-adiabatic temperature changes is a method leading to high performance, fast side-impact protection. In this contribution, we present a micromachined silicon temperature sensor for the detection of side-impacts in vehicles.

1 Introduction

Although the security of vehicle passengers in a crash incident has been much improved by means of passive security measures during the last years, further improvement of the passenger security is an intention of the car manufacturers. Especially optimisation of side impact protection is one objective, as the compression zone in side impacts is much smaller than in front or back impacts. Early and reliable detection of the impact is seen as a measure to achieve this goal. Today, side impact sensing is mostly realised using central acceleration sensors located between driver and co-driver or peripheral acceleration sensors near to potential impact positions. A new concept is the supplementation of these acceleration sensors with volume surveillance sensors [1]. With such sensors, the state of the air volume enclosed in the vehicle doors is kept under surveillance. Possible sensor types are air-pressure and air-temperature sensors. This paper focuses on the measurement of quasi-adiabatic changes of the air-temperature in the interior of the door due to a side-impact. For this measurement, extremely fast temperature sensors with response times of

3.5.6 Thermal Microsensor for the Detection of Side-Impacts in Vehicles

a few milliseconds are needed. Micromachining techniques, offer the possibility to realise silicon sensors with such a thermal dynamic.

2 Physical Background

In general, the interior of a vehicle side door can be seen as an air-filled cavity. Apart from minor leaks, this cavity is well sealed. If an intruding object deforms the walls of the cavity, the initial volume V_0 of the door interior will change by ΔV. If the intrusion occurs rapidly as is the case in a side impact, the enclosed air performs a change of state. In the ideal case (no leakage, no heat flow to the surroundings), this will be an adiabatic change leading to an air-temperature rise ΔT (Eq.1 left). As the change of the air-temperature will be small compared to the ambient temperature, it is possible to linearize the equation. This yields the right part of equation 1.

$$\Delta T = T_0 \left[\left(1 - \frac{\Delta V}{V_0}\right)^{1-\chi} - 1 \right] \xrightarrow{Lin} \Delta T = \frac{\Delta V}{V_0}(\chi - 1)T_0 \qquad \text{Eq. 1}$$

T_0: Initial temperature [K]
V_0: Initial door volume [m³]
ΔT: Rise of the air-temperature under ideal conditions [K]
ΔV: Reduction of the door volume [m³]
χ: Ratio of specific heats (1.402 for air)

For a typical side-impact case, an initial door volume of $V_0 = 0.05$ m³ and a volume reduction of $\Delta V = 0.005$ m³ during the impact can be presumed. If the impact occurs at an ambient temperature of $T_0 = 300$ K, this results in an air-temperature increase of $\Delta T = 12$ K. In reality, the door is not completely thermally isolated and airtight. Thus, the change of state of the enclosed air will only be quasi-adiabatic which results in a temperature change somewhat lower than calculated by equation 1. First measurements under real conditions showed temperature rises that account for 20% to 50% of that in the ideal case.

3 Sensor Concept

From section 2 follows the need for a sensor capable of sensing fast temperature changes ΔT as well as slow changes of T_0. Silicon micromachining offers the opportunity to realise temperature sensors with very short response times and thus, sufficient thermal dynamic for side-impact sensing [2]. A silicon temperature-sensor chip for the measurement of fast air-temperature changes has been developed at Robert Bosch GmbH. It is shown in Fig. 1.

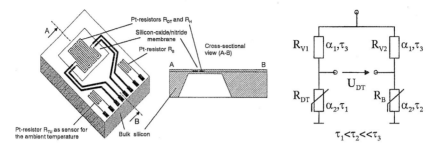

Fig. 1. Silicon Sensor Chip for the Measurement of Fast Temperature Changes

Fig. 2. Wheat Stone Bridge Sensitive to Fast Temperature Changes

The chip is realised in bulk micromachining technology. It comprises two temperature dependent platinum resistors R_{TU} and R_B, located on the 440 µm thick bulk silicon. These resistors react slowly to ambient temperature changes. On a 1.3 µm thick dielectric membrane, two further platinum resistors R_{DT} and R_H are located. The small thermal capacity of the membrane and its ability to thermally isolate the resistors from the bulk silicon results in a short response time of the resistors to air-temperature changes. For the measurement of fast temperature changes ΔT, a high-pass filter suppressing ambient temperature changes is needed. It is realised by a temperature sensitive wheat stone bridge shown in Fig. 2.

The upper part of the bridge consists of two discrete, external resistors R_{V1} and R_{V2}, having matched temperature coefficients α_1 and a thermal time constant τ_3, which is higher than that of the resistors on the sensor chip. The lower part of the bridge is built by R_{DT} and R_B, which are located on the sensor chip. Both possess the same matched temperature coefficient α_2. The thermal time constant τ_1 of R_{DT} is lower than the time constant τ_2 of R_B as it is located on the membrane. R_{V1} and R_{V2} are chosen in a way that the bridge stays balanced as long as the ambient temperature changes slowly. If the ambient temperature changes rapidly, the different time constants of R_{DT} and R_B yield an unbalanced bridge. Thus, U_{DT} shows a voltage peak corresponding to the temperature peak ΔT. The resistor R_H is used to realise a sensor self-test. By application of a current pulse to R_H the temperature of the membrane can be changed rapidly which results in a ΔT signal. Thus, the complete sensing path of the impact-sensor can be tested. The resistor R_{TU} is used for the measurement of the ambient temperature.

3.5.6 Thermal Microsensor for the Detection of Side-Impacts in Vehicles 319

4 Measurement Results

Based on the sensor concept described above, first thermal side-impact sensor prototypes have been realised. These were first investigated in a piston device capable of generating rapid volume changes. A result of this investigation is shown in Fig. 3.

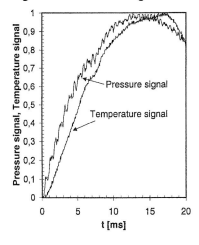 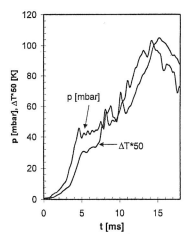

Fig. 3. Scaled Signals of a Pressure Sensor and a Thermal Side-impact Sensor in the Piston Device

Fig. 4. Temperature and Pressure Signals from a Crash Test

Then, the prototypes were tested in side-crashes. A result of a 29 km/h barrier-crash test is shown in Fig. 4. It can be seen that the thermal sensor shows the same signal shape as the pressure sensor. A short time delay ($\approx$ 1.5 ms) between the pressure sensor and the thermal sensor exists due to the remaining heat capacity of the dielectric membrane. It is expected that this time delay can be minimised by optimisation of the thermal sensor geometry.

5 Series Product

In a series product, the sensor electronic would be realised as an application specific integrated circuit. This will allow miniaturization of the sensor. The ASIC as well as the sensor chip will be mounted on a printed circuit board which will be enclosed in a small plastic housing capable for automotive applications. The size of the housing will be 55 mm x 25 mm x 18 mm. Suitable automotive connectors can be attached to the housing. Fig. 5 shows the proposed layout of the sensor series product.

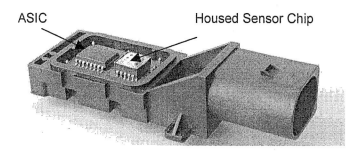

Fig. 5. Plastic Housing of the Thermal Side-impact Sensor. The housing lid has been opened to uncover the sensor chip and the ASIC.

6 Summary and Conclusion

In this contribution, a new micromachined thermal sensor for the detection of the quasi-adiabatic air-temperature increase in the case of a side crash has been presented. Part of the sensor concept is a temperature sensitive wheat stone bridge that is used to high-pass filter the temperature signal. Laboratory data as well as crash test data support the feasibility of this sensor concept.

References

[1] R. Aidam, P.Kocher R.-J. Recknagel: Studie: Alternative Seitencrashsensoren; VDI-Berichte Nr. 1637, 2001, p 185-197.

[2] I. Simon, M. Arndt: Thermal and Gas Sensing Properties of a Micromachined Thermal Conductivity Sensor. Transducers '01/Eurosensors XV, Munich, Germany, June 10-14, 2001, Digest of Technical Papers, p. 1492-1495.

3.5.7 Low-cost Filling Level Sensor

H. Ploechinger

E-mail: ploechinger@aon.at, info@thyracont.de
URL: www.ploechinger.com

Keywords: level sensor

Abstract

A new Filling Level Sensor is based on a 5 pole flat tape cable from industrial production and an especially developed switched-capacitor-ASIC. The Sensor can be used in a tank and it can be attached to the outer wall in case of plastic tanks.

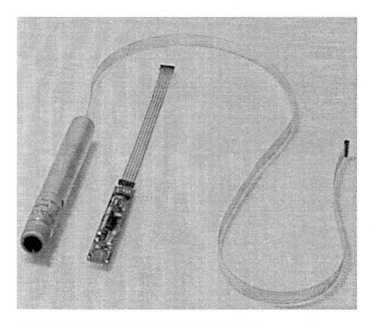

Fig. 1. Low-cost Filling Level Sensor

1 Introduction

Sensors without moving parts have established for level measurements in the industrial field. However, large amounts of sliding contact resistors are commonly used as level indicator in the automobile industry. Obviously for economical reasons developers and engineers still fall back on these mechanical, susceptible and rather inaccurate measuring devices.
Here an alternative is introduced, which is insensitive against wear and tear, easy to integrate in tanks of different types and shapes and which, above all, promises low system costs.

2 Homogeneous Media (Water, Oil, Braking Fluid, etc.)

The simplest type of this new sensor with capacitive measurement principle is based on a 5pole flat tape cable from industrial serial production with Kapton or Polyester insulation. The flat tape is connected to a small circuit board with an especially developed switched-capacitor-ASIC with differential path evaluation. This method of signal processing implies high noise immunity in principal and moreover requires no reference to ground.

If only one end of the cable is connected, the other is to be sealed, whereas this is not necessary if both ends are connected to the board as a loop. Both types can be attached to the inner tank wall as well as to the outer wall in case of plastic tanks.

Furthermore there is the possibility to imprint the 5 electrodes (2 pairs of capacitance electrodes with 1 separation electrode) on the tank walls as conductive film. Accordingly to the invention a transformation, e.g. linearisation, of the output characteristic can be achieved by varying the distance between the electrodes. By attaching the electrodes to two or even 4 sides of a symmetric tank inclination effects are compensated already in the sensor.

Operating the sensor with 5 V voltage supply yields an output from 1-4 V according to the filling level.

In it's basic version the sensor signal is depending on the fluid's constant of dielectricity. Hence, this version is suitable only for homogeneous media or mixtures of constant composition. A possibility of automatically considering the type of filling media is to attach a second flat tape cable as reference in a tank area, which is already covered by the fluid at low filling level.

3 Seat Occupation and Head Rest Positioning

The sensor with 5-electrode arrangement as described above can also be applied for recognition of seat occupation and correct head rest positioning in automobiles. First prototypes have been assembled and successfully tested.

4 Fuel Tank

Varying fuel composition, condensed water and the possibility of salt formation in the tank with uneven layers can seriously affect the accuracy of capacitive level measurements. The chance of accurate and reproducible measurements for these applications is offered by the following measuring principle, which is redundant to the above:

The separation electrode in the middle of the flat tape is realized as measurement loop. By means of a new thermodynamic measurement principle a signal corresponding to the filling level is generated by evaluating the thermal conductivity of the fuel and the gas mixture above.
The signals of the thermal and capacitive measurements are processed by a microcontroller, which evaluates these signals using a family of characteristic curves. The thermal measurement also provides a temperature signal.

Consequently this method results in an accurate level signal, which is far-reaching independent of the fluid's composition and temperature.

Advantages of this new concept:

- inexpensive, flexible sensor element
- high resolution, not limited by sensor principle
- high accuracy
- suitable for difficult locations and arbitrarily shaped tanks
- no wear and tear
- ASIC available
- high noise immunity
- cost-effective system

Additional advantages achieved by the version with a second thermal measurement:

- independent of fluid composition
- independent of the gas atmosphere over the fluid
- low-cost evaluation
- ability of self-testing
- redundance.

References

[1] patents DE 19516809 C1, DE 19645970 C2, WO 96/35929, US 6,178,818 B1, DE 10115715.0 Pat pend.

3.5.8 IBA - Integrated Bus Compatible Initiator

T. Goernig

Conti Temic microelectronic GmbH
Ringlerstrasse 17, 85057 Ingolstadt, Germany
Phone: +49/841/881-2493, Fax: +49/841/881-2420
E-mail: thomas.goernig@temic.com, URL: www.temic.com

Keywords: IBA, HfHx, initiator, integration, reliability, firing and sensor bus system

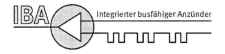

Abstract

This paper is focused on the cooperation project IBA (Integrated Bus compatible Initiator) and shall give an overview about the current status of the project. The project was initiated January 1st 2000 and is planned to be finalized mid 2002. The involved partners of the cooperation project IBA are the following companies and institutes: Dynamit Nobel GmbH, Fuerth; NICO Pyrotechnik Hanns-Juergen Diederichs GmbH & Co. KG, Trittau; Chemnitz University of Technology, Department ET/IT, Center of Microtechnologies, Chemnitz; Conti Temic microelectronics GmbH, Ingolstadt and TRW - Airbag Systems GmbH & Co. KG, Aschau/Inn. The research project is being funded by the BMFT in cooperation with the VDI/VDE-IT.

1 Introduction

With the development of standards for automotive passenger restraint firing and sensor bus systems, the need for integrated bus compatible initiators has became evident. The firing bus systems like the BoTe System from Bosch and Temic (three pin initiators, Daisy Chain architecture) or the SURFS System from Siemens (two pin initiators; parallel architecture) and the B/S/T firing and sensor bus system by Bosch, Siemens and Temic (combined Daisy Chain and parallel architecture) [1] are to be mentioned in this respect. An advantage of the bus systems is the increased flexibility as they can be introduced on a new carline by adding the additional components and reconfiguration of the bus master

[2]. An important component for a firing bus system is the integrated bus compatible initiator.

2   Integrated Bus Compatible Initiator

For the functionality of a restraint system based on the bus technology, it is important that the initiators of the various restraint devices have the required bus interface and the necessary electrical components included. The initiator developed within the IBA [3] project fulfils all the mechanical and electrical requirements:

- three or two connector pins for the bus interface (depending if parallel or serial bus architecture is selected)
- integration of all electrical components into the initiator:
    - bus interface including daisy chain switch
    - ESD protection (25 kV @ 150 pF;150 Ω)
    - energy reserve for electronics
    - firing energy capacitor
    - firing stages
    - safe and arm feature
    - bus protocol decoder and diagnostic circuit
- compatible with the mechanical dimensions of a conventional bridge-wire initiator
- environmental and life time requirements (e.g. $-40°C$ to $+90°C$; 15 years)

Due to assembly and technology restrictions the first generation IBA (IZE) had been built as a two-chamber design [4], [6]. The electrical components and pyrotechnics was encapsulated into two separate hermetically sealed chambers. Further focus of the IBA project is now on the consequent integration of the electrical components and pyrotechnics into one hermetical sealed chamber.

3   Initiator Bridge

The IBA project utilizes an earlier development of an initiator bridge that has been conducted within the SIFA/MST project. The HfHx initiator bridge has been further optimized and the structure of the element has been defined for further test series. The size of the HfHx silicon is app. 2 x 2 mm² [see Fig. 1] while the actual initiator bridge element is much smaller. Tests which have been conducted within the IBA project showed an energy consumption < 20 µJ. The initiator bridge is manufactured by using standard semiconductor processes [5].

Fig. 1.   HfHx-Initiator Bridge

4   Firing Energy and Test

To achieve the smallest possible mechanical package for the IBA it is important that the firing capacitor also has the smallest possible mechanical dimension. The required energy that is stored in the capacitor to reliably initiate the initiator bridge has been defined in various test series.

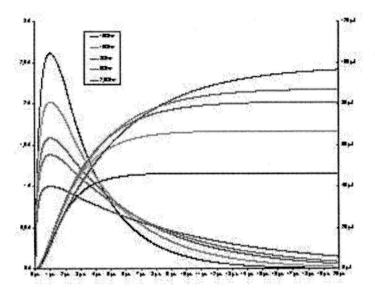

Fig. 2.   Typical Current and Energy at Initiator Bridge

The different parameters that have an influence on the result have been taken into account. Typical test curves can be seen in Fig. 2, an example of the activation of the HfHx initiator element is shown in Fig. 3. The parameters for the test series are defined as: C = 500 nF; ESR = 0,3 Ohm; Ron = 3 Ohm @ 2 µs; V(Bias) = 22,1 V.

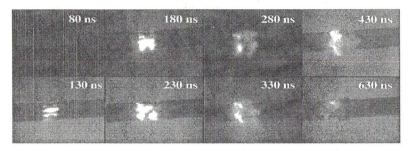

Fig. 3. Firing Test

5 IBA Initiator Assembly

The further project will also address manufacturing issues related to the IBA. This includes the highest integration, best usage of available and/or to be developed processes and optimum design for manufacturability. An example of the next generation IBA with a HfHx initiator chip is shown in Fig. 4.

Fig. 4. IBA Electronics with HfHx Initiator

6 Conclusion

With the results that are achieved within the IBA project, it will be possible to design and manufacture an integrated bus compatible initiator at economical costs. The consequent usage of all possibilities for miniaturization and the identification of additional synergy effects will lead to further cost optimized solutions and to a wide introduction of integrated bus compatible initiator. Small mechanical dimensions are an important issue for the integration of the IBA into the next generation inflators as shown in Fig. 5.

Fig. 5. IBA Initiator Design Study

Acknowledgements

The author like to thank the involved partners of the IBA project: G. Kordel (Dynamit Nobel GmbH, Fuerth), W. Schmid (NICO Pyrotechnik Hanns-Juergen Diederichs GmbH & Co. KG, Trittau), U. Weiss (NICO Pyrotechnik Hanns-Jürgen Diederichs GmbH & Co. KG, Trittau), Dr. B. Loebner (Chemnitz University of Technology, Department ET/IT, Center of Microtechnologies), Dr. H. Laucht (TRW - Airbag Systems GmbH & Co. KG, Aschau/Inn).

References

[1] Common Bosch Siemens Temic Bus Description Rev.: 2.0, Apr. 24th 2000, www.temic.com

[2] K. Balzer, C. Zelger, T. Goernig, BST Deployment and Sensor Bus, Airbag 2000+, 5th International Symposium on Sophisticated Car Occupant Safety Systems, Karlsruhe, Germany, Dec. 4-6, 2000.

[3] G. Kordel, W. Schmid, Dr. B. Loebner, Dr. H. Laucht, T. Goernig, IBA Status Seminar, Ingolstadt, Germany, Nov. 27th 2000.

[4] T. Goernig, Integrated Electronics for Bus System Igniters, mstnews 1/01, February 2001, www.vdivde-it.de/mstnews

[5] T. Goernig, Integrated Electronics for Bus System Igniters, AMAA 2001, Berlin, Germany, May 21-22, 2001, www.amaa.de

[6] S. Krueger, W. Gessner, AMAA Compendium 2001, www.amaa.de

Contributors

Ansorge, Frank
Fraunhofer Institute Reliability and Microintegration
Phone: +49 8153 9097-500
Fax: +49 8153 9097-511
Email: ansorge@mmz.izm.fhg.de

Anzinger, Claus
EADS Deutschland GmbH
Phone: +49 89 607-29969
Fax: +49 89 607-24001
Email: claus.anzinger@eads.net

Arndt, Michael
Robert Bosch GmbH
Phone: +49 7121 35-6343
Fax: +49 7121 35-376343
Email: michael.arndt@de.bosch.com

Becker, Joerg
Philips Semiconductors GmbH
Phone: +49 40 5613-3032
Fax: +49 40 5613-3045
Email: joerg.becker@philips.com

Büttner, Christof
Raytheon Deutschland GmbH
Phone: +49 8161 902-101
Fax: +49 8161 902-110
Email: c-buettner@raytheon.com

Carta, Giovanni
Infineon Technologies AG
Phone: +49 89 234-21537
Fax: +49 89 234-21237
Email: giovanni.carta@infineon.com

Cavalloni, Claudio
Kistler Instrumente AG
Phone: +41 52 224-1111
Fax: +41 52 224-1414
Email: claudio.cavalloni@kistler.com

Demeûs, Laurent
CISSOID S.A.
Phone: +32 10 48 92 10
Fax: +32 10 48 92 19
Email: laurent.demeus@cissoid.com

Contributors

Dordet, Yves
Siemens VDO Automotive S.A.S.
Phone: +33 5 61 19 72 94
Fax: +33 5 61 19 25 35
Email: yves.dordet@at.siemens.fr

Gaißer, Alexander
Hahn-Schickard-Gesellschaft
Institute of Micromachining
and Information Technology
Phone: +49 7721 943-226
Fax: +49 7721 943-210
Email: alexander.gaisser@hsg-imit.de

Gessner, Wolfgang
VDI/VDE-IT
Phone: +49 3328 435-173
Fax: +49 3328 435-225
Email: gessner@vdivde-it.de

Goernig, Thomas
Conti Temic microelectronic GmbH
Phone: +49 841 881-2493
Fax: +49 841 881-2420
Email: thomas.goernig@temic.com

Grace, Roger
Roger Grace Associates
Phone: +1 415 436 9101
Fax: +1 415 436 9810
Email: rgrace@rgrace.com

Grüger, Heinrich
Fraunhofer Institut für Mikroelektronische
Schaltungen und Systeme
Phone: +49 351 8823-155
Fax: +49 351 8823-166
Email: heinrich.grueger@imsdd.fhg.de

Kumar, Bal
BAE Systems Advanced Technology Centre
Phone: +44 20 8624 6586
Fax: +44 20 8624 6099
Email: bal.kumar@baesystems.com

Krueger, Sven
VDI/VDE-IT
Phone: +49 3328 435-221
Fax: +49 3328 435-256
Email: krueger@vdivde-it.de

Lages, Ulrich
IBEO Automobile Sensor GmbH
Phone: +49 40 64587-170
Fax: +49 40 64587-109
Email: ul@ibeo.de

Langheim, Jochen　　CARSENSE Consortium
Phone: +33 2984 459443
Fax:　　+33 2984 95655
Email: jochen.langheim@autocruise.fr

Moulin, Philippe　　Ricardo Consulting Engineers Ltd.
Email: PMoulin@ricardo.com

Pearce, Kate　　Emerson & Cuming
Phone: +1 978 436-9700
Fax:　　+1 978 436-9707
Email: kate.l.pearce@nstarch.com

Plöchinger, Heinz　　Thyracont GmbH
Phone: +49 851 95986-0
Fax:　　+49 851 95986-40
Email: heinz.ploechinger@thyracont.de

Reibe, Thomas　　European Commission
Information Society Directorate-General
Phone: +32 2 295 61 25
Fax:　　+32 2 299 82 49
Email: thomas.reibe@cec.eu.int

Reischl, Stefan　　Infineon Technologies AG
Phone: +49 89 234-20345
Fax:　　+49 89 234-716418
Email: stefan.reischl@ infineon.com

Schmid, Ulrich　　EADS Deutschland GmbH
Phone: +49 89 607-20657
Fax:　　+49 89 607-24001
Email: ulrich.schmid@eads.net

Schmidt, Rainer　　Fuji Electric GmbH
Phone: +49 69 669029-22
Fax:　　+49 69 6690 29-56
Email: rschmidt@fujielectric.de

Seegebrecht, Peter　　Christian-Albrechts-Universität zu Kiel
Phone: +494318806075
Fax:　　+494318806077
Email: ps@tf.uni-kiel.de

Teepe, Gerd　　Motorola GmbH
Phone: +49 89 92103-880
Fax:　　+49 89 92103-101
Email: gerd.teepe@motorola.com

Teuner, Andreas Delphi Delco Electronics Europe GmbH
Phone: +49 202 291-3410
Fax: +49 202 291-4140
Email: andreas.teuner@delphiauto.com

Tissot, Jean-Luc CEA / LET - DOPT
Phone: +33 4 38 78 40 91
Fax: +33 4 38 78 51 73
Email: jltissot@cea.fr

Tomasi, Luca EADS Deutschland GmbH
Phone: +49 89 607-24029
Fax: +49 89 607-24001
Email: luca.tomasi@eads.net

van Dommelen, Ignas European Semiconductor Assembly (Eurasem) BV
Phone: +31 24 371 44 97
Fax: +31 24 377 04 06
Email: i.vandommelen@eurasem.com

Wallin, Christer ABB Automation Technology Products AB
Phone: +46 21 342498
Email: christer.s.wallin@se.abb.com

Weiss, Elmar BMW AG
Phone: +49 89 382-30766
Fax: +49 89 382-46520
Email: elmar.weiss@bmw.de

Wicht, Henning WTC-Wicht Technology Consulting
Phone: +49 89 9280428
Fax: +49 89 9280455
Email: henning.wicht@wtc-consult.de

List of Keywords   Page

Actuation force	251
Actuator	164
Adaptive cruise control	116
Angle sensing	227
Angular position detection	222
Angular rate sensor	311
Anti-blocking-system	243
Anticipatory crash-sensing	89
Application	157
Application status	3
Assembly	289
Automotive trends	13
Brake-by-wire	251
Brake pedal	251
Bus topology	276
Calibration after assembled	271
CAN	276
Chalcogenide	141
Challenges	3, 13
Circuit protection	297
Combustion	196
Common rail	174
Contactless	212
Cramshaft	206
Diaphragm	251
Digital trimming	271
Direct injection	164
Double diaphragm package	271
Driver assistance	89, 96
Electronic beam scanning	116
Electro-optical systems	89
Encapsulant	297

Faraday rotation	116
Ferrite	116
Ferroelectric	121
Firing and sensor bus system	325
FlexRay	276
Fluorosilicone	297
Focal plane array	141
Fusion	96
Gasoline	164
Gyroscope	311
Hall	227
Hall sensors	206, 222
Harsh environment	306
HfHx	325
High pressure	174
High temperature sensors	306
Hot film anemometer	174
IBA	106, 325
Initiator	325
Injection rate	174
Injection system	174
Injector	164
Integration	106, 325
IR lens	141
Laser	96
Laser scanner	136
Level sensor	321
LIDAR	89
LIN	276
Linear	227
Low power	306
Magnetic saturation	212
Magnetism	222
Magnetostrictive sensor	184
Market analysis	3
MEMS-packaging	146
Microbolometer	141
Micro-mechatronics	146

Micro-mirrors	146
Millimetric wave lengths	116
Mounting	136
Multiplexing	276
Nearfield	136
Networking	276
NEXUS	13
Night vision	121
Non-contact	184
Non-uniformity correction	121
Object outline	136
Obstacle	136
Occupant position recognition	89
On-board-diagnosis	174
Optical sensors	289
P2SC	243
Pedal position potentiometer replacement	227
Piezoelectric sensor	164
Piezoeresistive sensors	196
Plastic packaging	289
PMD	106
Position sensor	212
Powertrain	157, 184
Pre crash sensing	106
Pressure sensing	196
Pressure sensor	157, 232, 243, 252, 261
Printed coil	212
Products	3, 13
RADAR	89, 96
Regulation	26
Reliability	212, 306, 325
Safety	106, 212
Safety systems	26
Self-test	261
Sensing 360°	106
Sensors	276, 306
Side-airbag	316
Side-impact	316
Silicon on insulator	196
Silicone	297
Single chip	222

SOI	306
SOI-technology	311
Standardisation	251
Standardised technology	212
STARC	243
Stroke amplifying unit	164
Thermal microsensor	316
Tire sensing	243
Torque	184
TPMS	243
Tracking	136
Trends	3
True power on	206
Twisted independent mounting	206
Two levels sensitivity	261
Uncooled infrared sensors	121, 141
Video	96
Video processing	121

Printing: Mercedes-Druck, Berlin
Binding: Buchbinderei Lüderitz & Bauer, Berlin